T0184057

Lecture Notes in Artificial Intelligence 10733

Subseries of Lecture Notes in Computer Science

LNAI Series Editors

Randy Goebel
University of Alberta, Edmonton, Canada
Yuzuru Tanaka
Hokkaido University, Sapporo, Japan
Wolfgang Wahlster
DFKI and Saarland University, Saarbrücken, Germany

LNAI Founding Series Editor

Joerg Siekmann
DFKI and Saarland University, Saarbrücken, Germany

More information about this series at http://www.springer.com/series/1244

Qiang Fang · Jianwu Dang
Pascal Perrier · Jianguo Wei
Longbiao Wang · Nan Yan (Eds.)

Studies on Speech Production

11th International Seminar, ISSP 2017
Tianjin, China, October 16–19, 2017
Revised Selected Papers

 Springer

Editors
Qiang Fang
Chinese Academy of Social Sciences
Beijing
China

Jianwu Dang
JAIST
Nomi
Japan

Pascal Perrier
Grenoble Alpes University
Saint-Martin-d'Hères
France

Jianguo Wei
Tianjin University
Tianjin
China

Longbiao Wang
Tianjin University
Tianjin
China

Nan Yan
Shenzhen Institute of Advanced Technology
Shenzhen
China

ISSN 0302-9743   ISSN 1611-3349   (electronic)
Lecture Notes in Artificial Intelligence
ISBN 978-3-030-00125-4   ISBN 978-3-030-00126-1   (eBook)
https://doi.org/10.1007/978-3-030-00126-1

Library of Congress Control Number: 2018953363

LNCS Sublibrary: SL7 – Artificial Intelligence

© Springer Nature Switzerland AG 2018
This work is subject to copyright. All rights are reserved by the Publisher, whether the whole or part of the material is concerned, specifically the rights of translation, reprinting, reuse of illustrations, recitation, broadcasting, reproduction on microfilms or in any other physical way, and transmission or information storage and retrieval, electronic adaptation, computer software, or by similar or dissimilar methodology now known or hereafter developed.
The use of general descriptive names, registered names, trademarks, service marks, etc. in this publication does not imply, even in the absence of a specific statement, that such names are exempt from the relevant protective laws and regulations and therefore free for general use.
The publisher, the authors and the editors are safe to assume that the advice and information in this book are believed to be true and accurate at the date of publication. Neither the publisher nor the authors or the editors give a warranty, express or implied, with respect to the material contained herein or for any errors or omissions that may have been made. The publisher remains neutral with regard to jurisdictional claims in published maps and institutional affiliations.

This Springer imprint is published by the registered company Springer Nature Switzerland AG
The registered company address is: Gewerbestrasse 11, 6330 Cham, Switzerland

Preface

We are pleased to present the proceedings of the 11th International Seminar on Speech Production (ISSP 2017). This conference was co-organized by Tianjin High-Tech Industry Association, Tianjin University, and the Institute of Linguistics of the Chinese Academy of Social Sciences (CASS) and held during October 16–19 in Tianjin, China.

The conference was aimed to bring together leading academic scientists, researchers, and scholars to exchange and share their experience and research results on all aspects of speech production and to discuss practical challenges encountered and the solutions adopted. 68 presentations were contributed to ISSP 2017, covering a wide range of speech science fields including phonology, phonetics, prosody, mechanics, acoustics, physiology, motor control, neuroscience, computer science, and human interaction. 89 participants came from 11 countries, namely: Australia, Belgium, Canada, China, France, Germany, Hong Kong, Hungary, Japan, UK, and USA. All presentations were in English.

The 20 papers in this issue were selected from 68 presentations (acceptance rate of 29.4%) accepted by ISSP 2017, which have been peer-reviewed by international experts from the conference technical committee. These leading studies will give excellent presentations that are anticipated to break through the barriers separating various disciplines among the various fields of speech science. Finally, our appreciation is extended to all the participants who contributed in making this conference successful and beneficial.

July 2018

Qiang Fang
Jianwu Dang
Pascal Perrier
Jianguo Wei
Longbiao Wang
Nan Yan

Organization

Organizing Committee

General Conference Chairs

Jianwu Dang	Tianjin University, China
Aijun Li	Chinese Academy of Social Sciences, China

Technical Program Chairs

Pascal Perrier	GIPSA-Lab, France
Qiang Fang	Chinese Academy of Social Sciences, China
Nan Yan	Chinese Academy of Sciences, China
Longbao Wang	Tianjin University, China

Local Arrangement Chair

Jianguo Wei	Tianjin University, China

Special Session Chairs

Kiyoshi Honda	Tianjin University, China
Xiyu Wu	Peking University, China
Ming Li	Duke Kunshan University, China

Tutorial Chairs

Phil Hoole	Ludwig-Maximilians-Universität München, Germany
Fang Hu	Chinese Academy of Social Sciences, China
Jinsong Zhang	Beijing Language and Culture University, China

Plenary Session Chairs

Mark Tiede	Haskins Laboratories, USA
Liang Ma	Fudan University, China

Publication Chairs

Jianhua Tao	Chinese Academy of Sciences, China
Wentao Gu	Nanjing Normal University, China
Yu Chen	Tianjin University of Technology, China

Publicity Chairs

Tatsuya Kitamura	Konan University, Japan
Jia Jia	Tsinghua University, China

Jiahong Yuan   Linguistic Data Consortium and University
                          of Pennsylvania, USA
Qingzhi Hou   Tianjin University, China

Registration Chairs

Hui Feng   Tianjin University, China
Xin Dang   Tianjin Polytech University, China

Additional Reviewers

Ya Li   Chinese Academy of Sciences, China
Xudong Zheng   University of Maine, USA
Véronique Delvaux   Université de Mons, Belgium
Amelie Rochet-Capellan   GIPSA-lab, France
Johanna-Pascale Roy   Université Laval, Canada
Benjamin Swets   Grand Valley State University, USA
Paavo Alku   Aalto University, Finland
Véronique Boulenger   Université de Lyon, France
Yuan Jia   Chinese Academy of Social Sciences, China
Heidi Altmann   Universität Stuttgart, Germany
Maria Paola Bissiri   Queen Margaret University, UK
Si Chen   The Hong Kong Polytechnic University, China
Qian Xue   University of Maine, USA
Tatsuya Kitamura   Konan University, Japan
Sabine Skodda   Universitätsklinikum Knappschaftskrankenhaus Bochum,
                          Germany
Albert Rilliard   LIMSI-CNRS, France
Brad Stoy   University of Arizona, USA
Christine Mooshammer   Haskins Laboratories, USA
Christopher Carignan   Western Sydney University, Australia
Feng-fan Hsieh   National Tsing Hua University, Taiwan
Longbiao Wang   Tianjin University, China
Lucie Ménard   Université du Québec à Montréal, Canada
Rudolph Sock   Université de Strasbourg, France
Wentao Gu   Nanjing Normal University, China
Phil Hoole   Ludwig-Maximilians-Universität München, Germany
Pierre Badin   GIPSA-lab, France
Aijun Li   Chinese Academy of Social Sciences, China
Mark Tiede   Haskins Laboratories, USA
Liang Ma   Fudan University, China
Peggy Mok   The Chinese University of Hong Kong, SAR China
Susanne Fuchs   Centre for General Linguistics, Germany
Chakir Zeroual   Université Sidi Mohamed Ben Abdellah, Morocco
Xiyu Wu   Peking University, China
Kiyoshi Honda   Tianjin University, China

Mengxue Cao	Beijing Normal University, China
Christin Shadle	Haskins Laboratories, USA
Nathalie Henrich	French National Centre for Scientific Research, France
Feng Ling	Shanghai University, China
Fang Hu	Chinese Academy of Social Sciences, China
Wai-Sum Lee	City University of Hongkong, SAR China
Pascal Perrier	GIPSA-lab, France
Nan Yan	Chinese Academy of Sciences, China
Rushen Shi	Université du Québec à Montréal, Canada
Jean-Luc Schwartz	GIPSA-lab, France
Haibo Wang	Chinese Academy of Social Sciences, China
Peter Birkholz	Technische Universität Dresden, Germany
Christophe Savariaux	GIPSA-lab, France
Alan Wrench	Articulate Instruments Lt, UK
Hironolri Takemoto	Chiba Institute of Technology, Japan
Tsukasa Yoshinaga	Osaka University, Japan
Thomas Hueber	GIPSA-lab, France
Jianwu Dang	Tianjin University, China
Yohann Meynadier	Aix-Marseille Université, France
Zhenhua Ling	University of Science and Technology of China, China
Cécile Fougeron	Université Sorbonne Nouvelle - Paris 3, France
Doris Mücke	University of Cologne, Germany
Gérard Bailly	GIPSA-lab, France
Jana Brunner	Humboldt-Universität zu Berlin, Germany
Jangwon Kim	Canary Speech, USA
Sophie Dufour	Aix-Marseille Université, France
Nathalie Henrich	GIPSA-lab, France
Donna Erickson	Haskins Laboratories, USA
Xiaojie Zhao	Beijing Normal University, China
Mariane Pouplier	Ludwig-Maximilians-Universität München, Germany
Seiji Adachi	Gifu University, Japan
Caterina Petrone	Aix-Marseille Université, France
Marija Tabain	La Trobe University, Australia
Qiang Fang	Chinese Academy of Social Sciences, China
Doug Whalen	Haskins Laboratories, USA
Yinghao Li	Yanbian University, China
Alexandra Markó	Eötvös Loránd University, Hungary
Kathy Huet	Université de Mons, Belgium
Yu Chen	Tianjin University of Technology, China
Manwa Ng	The University of Hong Kong, SAR China
Na Zhi	Capital Normal University, China
Ao Chen	Beijing Language and Culture University, China
Jue Yu	Tongji University, China
Yinyi Luo	Chinese Academy of Social Sciences, China
Dengfeng Ke	Chinese Academy of Sciences, China
Yueh-chin Chang	National Tsing Hua University, Taiwan

Phil McAleer	University of Glasgow, UK
Sarah Hirschmüller	Johannes Gutenberg-Universität Mainz, Germany
Olov Engwall	KTH Royal Institute of Technology, Sweden
Jean Schoentgen	Université Libre de Bruxelles, Belgium
Yanmin Qian	Shanghai Jiao Tong University, China

Sponsor

Tianjin High-Tech Enterprise Association (THTEA)

Contents

Phonetics

Speech Planning and Comprehension

Speech Disorder

Emotional Speech Analysis and Recognition

Personality Judgments Based on Speaker's Social Affective Expressions

Donna Erickson[1,2,3(✉)], Albert Rilliard[4,5], João de Moraes[5], and Takaaki Shochi[6,7]

[1] Haskins Laboratories, Haven, USA
EricksonDonna2000@gmail.com
[2] Kanazawa Medical University, Ishikawa, Japan
[3] Sophia University, Tokyo, Japan
[4] LIMSI, CNRS, Université Paris-Saclay, Orsay, France
albert.rilliard@limsi.fr
[5] Laboratório de Fonética Acústica, FL/UFRJ/CNPq, Rio de Janeiro, Brazil
jamoraes3@gmail.com
[6] CLLE-ERRSaB, UMR 5263, Bordeaux, France
[7] LaBRI UMR 5800, Talence, France
takaaki.shochi@labri.fr

Abstract. This paper describes some of the acoustic characteristics that influence peoples' judgments about others. The database used was the multilingual corpus recorded with speakers in communicative dialogue contexts (e.g., Rilliard et al. 2013). The acoustic measurements were F0, intensity, HNR, H1-H2, and formant frequencies (F1, F2, and F3). The personality assessment was based on that proposed by Costa and McCrae (1992). A Multiple Factor Analysis (MFA) related the acoustic measures, the performance scores for each attitude, and the number of high, high-medium, low-medium and low ratings in the 5 personality traits, for audio-only and for audio-visual modalities. The results show that the most expressive speakers, those who produced the widest range in acoustic changes, were perceived as more EXTROVERTED and CONSCIENTIOUS. Speakers with high noise levels in the voice were judged with low AGREEABLENESS, and produced the best expressions involving an imposition on the interlocutor. Speakers judged as having high NEUROTICISM and low OPENNESS were perceived as the best performers for expressions with strong social constraints.

Keywords: Social affective expressions · Personality judgments
Acoustic analysis

1 Introduction

People make first impressions about a person—their age, social or ethnic origin, personality, current mood, emotional state, etc.—based on a variety of information (e.g., wwnorton.com/college/psych/personalitypuzzle6/ch/06/review.aspx), including facial features, physical attributes and appearances, but also their voice and speech characteristics (e.g. Scherer 1972; Lippa 1978; Ekman et al. 1980; Borkenau and Liebler

© Springer Nature Switzerland AG 2018
Q. Fang et al. (Eds.): ISSP 2017, LNAI 10733, pp. 3–13, 2018.
https://doi.org/10.1007/978-3-030-00126-1_1

1992). Voice cues, those that don't take part in the actual linguistic message (cf. the definition of "prosody" in Swerts and Kramer 2005) give information about a speaker's emotional state and expressivity (e.g., Goudbeek and Scherer 2010). Emotion-related cues are linked to physiological changes, and as such, similar across languages and cultures (e.g., Scherer et al. 2004). Conventional use of such cues, driven by symbolic and cultural choices (e.g., Léon 1993; Ohala 1994; Madureira and Camargo 2010) are used to facilitate social and interpersonal interactions, and it is these conventionalized cues in the voice which tend to vary across languages (Delattre 1963), as does the inventory of such conventional expressions (Shochi et al. 2009a, b). We refer to these conventionalized expressions as "social affective expressions" (Halberstadt et al. 2001). An interesting question is how do the acoustic changes in the speech signal used to express social affective expressions influence peoples' judgments about what kind of personality a person has. Work by Costa and McCrae (1992), and elaborated on by Nettle (2007), suggests that peoples' personalities can be described in terms of five major dimensions/traits along which all human characters vary. Traits stem from the way their nervous systems are wired up; they are defined as "stable individual differences in the reactivity of mental mechanisms designed to respond to particular classes of situations" (Nettle 2007: 43.). Each person has a certain amount of each of the five traits, and the combination of traits is what defines a personality. The traits can only be observed through a person's behavior, not directly. The five traits are referred to under the acronym, OCEAN (Table 1).

Table 1. 5-Trait OCEAN Model of Personality.

Trait	Description
OPENNESS	Creative, imaginative, eccentric vs. Practical, conventional
CONSCIENTIOUSNESS	Organized, self-directed vs. Spontaneous, careless
EXTROVERSION	Outgoing, enthusiastic vs. Aloof, quiet
AGREEABLENESS	Trusting, empathetic vs. Uncooperative, hostile
NEUROTICISM	Prone to stress & worry vs. Emotionally stable

In this paper we use the five trait OCEAN model to investigate (1) whether people can assess others' personalities from audio-visual recordings of a person uttering a variety of social affective expressions and (2) what are the acoustic cues associated with that personality type. The personality assessment is based on the OCEAN personality traits.

2   Method

2.1   Data Base Used

The database used was the multilingual audio-visual corpus recorded with speakers in communicative dialogue contexts, with set communication goals guided by an interlocutor and predefined hierarchical relationships (e.g., Rilliard et al. 2013).

The dialogues were set up to yield two target utterances ("Mary was dancing" and "a banana"), each produced with sixteen expressive variants.

This paper focuses on the 16 expressive variants of the utterance "a banana", as spoken by 19 speakers (8 U.S. English speakers (3 m, 5 f), 5 French speakers of English (3 m, 2 f) and 6 Japanese speakers of English (3 m, 3 f) (see Rilliard et al. 2013, 2014 for a detailed description of utterance selection). The 16 social affective expressions were selected as representative of a set of possible social expressions, but not inclusive of all possible ones (e.g. Uldall 1960). The expressions are labeled as follows: admiration, arrogance, authority, contempt, doubt, irony, irritation, neutral declarative sentence, neutral question, obviousness, politeness, seduction, sincerity, surprise, uncertainty, and "walking on eggs"—cf. Rilliard et al. (2013) for all definitions. Individual behavioral differences in the expression of these attitudes may reflect some aspects of the speaker's personality—the interpretation of these behavioral changes is the aim of the present work, not the actual performance on individual attitudes.

2.2 Performance Scores

Since speakers produce social affective expressions with a certain amount of variation, their performances were rated on a 1 to 9 scale by a total of 68 L1 USA English speakers as to how well they thought each speaker produced the intended expression (details in Rilliard et al. 2013, 2014). Quality scores given by each listener were standardized.

2.3 Acoustic Measurements

Six acoustic measurements were made: (1) F0, measured on each vowel using Praat's standard algorithm, and expressed in semitones relative to 1 Hertz (ST – measures were not corrected for individual or gender differences, as such difference in pitch is perceived by listeners, and a gender effect is observed in the perception results), (2) Intensity, A-weighted intensity measured on vowels using a Praat script, and expressed in decibels (dB) (note that the recordings were made using a calibrated microphone to correct for change in gain control, thus allowing comparison of speaker differences in their produced sound levels), (3) Harmonic-to-Noise Ratio (HNR), measured on vowels, expressed in dB, (4) Difference between the first two harmonics amplitude (H1-H2) using the COVAREP toolbox (Degottex et al. 2014), expressed in dB, (5 & 6) Measures of the first two formants using the Praat standard algorithm, with the parameters recommended for female or male voices (see Rilliard et al. 2016 for details). The median and range (measured as the difference between the 90th and 10th percentiles) values of these measurements over each sentence were taken as input of the analysis (totaling 12 measures: two on each acoustic parameter).

2.4 Personality Assessment

36 subjects, all first language speakers of English at a Midwestern university in the United States took the test; 15 were presented with the audio-only performances

(5 f, 10 m) and 21 (12 f, 9 m) with the audio-visual performances. The test was run on a computer with headphones, using an interface based on Runtime Revolution software.

Each subject was asked to listen to the performances of each of the 19 speakers uttering "a banana" with 16 different affects (listed above). The raw performances on these 16 attitudes were kept as any delexicalization process (e.g. Sonntag and Portele 1998) may alter voice quality and the subtle acoustic relations between audio and visual modalities, possibly impairing ratings; reproducing these results using prosody-only stimuli is interesting but challenging.

After listening to a speaker utter 16 variations of "a banana", ratings were done using the "Newcastle Personality Questionnaire" (Nettle 2007) containing 12 questions, such as "do you think this person would start a conversation with a stranger? Make sure others are comfortable and happy? Use difficult words?" etc. The ratings were on a 5-point scale ranging from "agree strongly" to "disagree strongly". The answers allow an estimation of the five personality traits proposed in Costa and McCrae (1992). See Table 2 for the twelve questions along with corresponding trait.

Table 2. The twelve questions composing the questionnaire subjects had to fill in for rating the perceived speakers' personality, with the trait each question contributes to rate (E: extraversion, N: neuroticism, C: conscientiousness, A: agreeableness, O: openness).

Assertion	Trait
Starting a conversation with a stranger	E
Making sure others are comfortable and happy	A
Creating an artwork, piece of writing, or piece of music	O
Preparing for things well in advance	C
Feeling blue or depressed	N
Planning parties or social events	E
Insulting people	A
Thinking about philosophical or spiritual questions	O
Letting things get into a mess	C
Feeling stressed or worried	N
Using difficult words	O
Sympathizing with others' feelings	A

The scores are then transformed into four categories (low, low-medium, high-medium, high) according to the spread of population on these scores, as described in Nettle (2007: 272ff).

2.5   Analysis of Acoustics, Performance Scores and Personality Traits

A Multiple Factor Analysis (MFA) (Bécue-Bertaut and Pagès Bécue-Bertaut and Pagès 2008, for mathematical details on the method; also, see R's FactoMineR library Husson et al. 2010) was applied so to relate the 12 acoustic measures, the performance scores for each attitude, and the number of high, high-medium, low-medium and low ratings in the 5 personality traits, for audio-only and for audio-visual modalities (5 scores for 5 traits in 2 modalities). An MFA performs multidimensional analyses (typically principal component analysis) on each table in a set of tables having common rows or observations and reporting different types of measures (here the three tables with respectively the acoustic measures, the performance scores, and the personality judgments—all measures summarizing one aspect of each speaker variation; the 19 speakers constitute the row of the matrices); the main dimensions of each analysis show the dispersion of the row in sets of Euclidean spaces with a given number of abstract dimensions. These dimensions are then matched by associating the position of the rows in each individual analysis during a general analysis matching rows over tables.

3 Results

For each personality trait, contingency tables were built on the basis of the subjects' answers, counting how many times each of the four levels of the scale was selected for each speaker. The tables are analyzed with a log-linear model allowing tile representations, as shown in Figs. 1, 2 and 3. The color of tiles indicates speakers who significantly depart (positively in blue or negatively in red) from the average of a given scale level for a personality trait, while the type of line around the tiles indicates if this departure is positive (plain lines–speakers who are more often perceived than average on a given level for a personality trait) or negative (broken lines–speakers who are less often perceived than average on a given level for a personality trait).

Figure 1 shows the results for EXTROVERSION: three speakers are judged with high levels of extroversion: S24, S3, S6. The first two are judged so in both modalities, S3 only in the AV modality. Four speakers (S12, S15, S25, S29) are judged with low

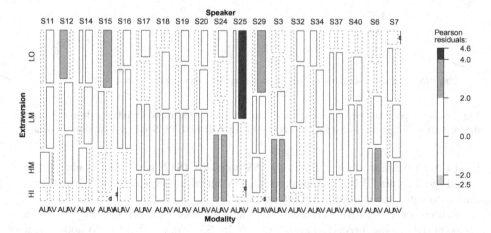

Fig. 1. Tile representation of contingency matrix for speakers in each modality (in columns) of the number of judgments in the four levels of EXTROVERSION (in lines: Low, Low-medium, High-medium, High). The levels significantly departing from an average distribution are shown in blue (for excess), or in red (for fewer ratings). (Color figure online)

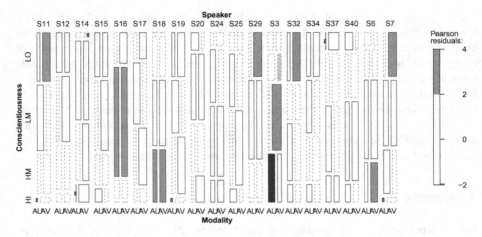

Fig. 2. Tile representation of contingency matrix for speakers in each modality (in columns) of the number of judgments in the four levels of CONSCIENTIOUSNESS. See Fig. 1 caption for details.

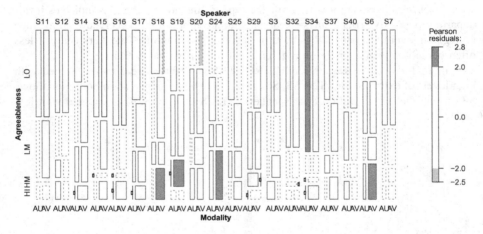

Fig. 3. Tile representation of contingency matrix for speakers in each modality (in columns) of the number of judgments in the four levels of AGREEABLENESS. See Fig. 1 caption for details.

level of extroversion–all but S12 in the AV modality (S12 in the audio only). Extroverted speakers are L1 English speakers; the AV introverted speakers are L2 speakers – thus a potential language/cultural effect.

Ratings of CONSCIENTIOUSNESS varied across speakers (cf. Fig. 2). Three speakers (L1 females) received high scores (S18, S3, S6), the first two in both modalities, S6 in AV only. Four speakers received low scores (S11, S29, S32, S7), all in the AV modality. All of them are male, and mostly L2 speakers. For AGREEABLENESS (cf. Fig. 3), four speakers (L1 females) received high or high-medium agreeableness scores (S18, S19, S24, S6); and one speaker (L2 male) was judged as having a disagreeable personality (S34). All the positive judgments were made in the AV modality, the negative one being done in the audio modality. For NEUROTICISM and OPENNESS, few tendencies emerge since the results are spread over the scale, and due to space, are not shown here.

In order to analyze further the changes in personality trait ratings made on the basis of observations of audio or audio-visual performances, PCA were run on each trait scale in each modality (complete results cannot be presented here, but see Rilliard et al. 2016 for more details). Most PCA have their first dimension correlated with the high-low dimension. For NEUROTICISM and OPENNESS, where the spread is large, many answers are grouped toward the lower bound, and the second dimension is used to differentiate the remaining high scores.

Results of a Multiple Factor Analysis (MFA) are detailed in Table 3 which shows the variables that are significantly associated to the first three dimensions of the MFA (that explain more than 40% of the variance). No significant associations are observed for the other dimensions. The MFA first dimension is positively associated to the perception of high extroversion, conscientiousness, and agreeableness, and negatively to low measures of extroversion and conscientiousness. This may be related to the importance of range of F0, F1 and intensity, and a high F0 register – which relates to good performance of irritation. The first dimension is related to the speaker's gender, with females having higher F0 register, and being rated as more extroverted than males.

The second dimension is inversely related to judgments of agreeableness and openness; it is also linked to voices with a large range of noise, large intensity variation, small F2 variations, and low median HNR. These perceptions are linked to good performances in authority, surprise, seduction, contempt and irritation. A relation with a speaker's gender is also observed (weaker than the one observed for the first dimension, but significant), with females having higher ratings than males on the side linked to high agreeableness, and less noisy voices.

The third dimension is correlated to the perception of neuroticism. It is correlated to the perception of high performances in sincerity, declaration, authority and "walking-on-eggs" – most expressions having strong social values and constraints. No specific relations with the measured acoustic variables have been found. For the two supplementary variables linked to speakers (gender and language background), only gender shows significant relation with the three sets of measures: there is thus no observed link between the speakers' linguistic origin and personality judgment by interlocutors. On the contrary, females in our set of speakers are rated with higher levels on the extroversion, conscientiousness, agreeableness and openness traits (no effect on neuroticism is observed). For agreeableness and openness, the video display may play a role, as most correlates are only observed within the audio-visual group (but for high agreeableness). The relatively limited number of speakers of both genders did not allow us to draw many conclusions on gender, given the complex nature of these observations.

Table 3. Measures of association (r – correlations for acoustic, performance and personality variables, regression coefficient for supplementary variable) between the variables and the first three dimensions of the MFA (negative correlations in italics). Personality measures are referred to according the first letter of the personality trait (E: extroversion, N: neuroticism, C: conscientiousness, A: agreeableness, O: openness), the modality (au: audio, av: audio-visual), and the ratings (high, high-medium, low-medium, low).

1st dimension		2nd dimension		3rd dimension	
Name	r	Name	r	Name	r
Acoustic					
F1 range	0.76	HNR range	0.72		
F0 range	0.64	INT range	0.59		
INT range	0.60				
F0 median	0.60				
		HNR median	*−0.70*		
		F2 range	*−0.70*		
Performance					
IRRI	0.59	AUTH	0.58	SINC	0.51
		SURP	0.58	DECL	0.48
		SEDU	0.56	AUTH	0.47
		CONT	0.46	WOEG	0.47
		IRRI	0.46		
Personality					
E, au, high	0.85	A, av, low	0.67	N, au, hi-med	0.69
C, av, high	0.82	E, av, hi-med	0.56	O, au, lo-med	0.66
C, au, high	0.71				
C, av, hi-med	0.71				
A, au, high	0.68				
E, av high	0.68				
E, av, low	*−0.56*	*A, av, lo-med*	*−0.57*	*N, au, low*	*−0.60*
C, au, lo-med	*−0.59*	*E, av, low*	*−0.61*		
C, av, lo-med	*−0.64*	*A, au, hi-med*	*−0.62*		
C, av, low	*−0.65*	*O, av, hi-med*	*−0.73*		
Supplementary variable					
Gender	0.53	Gender	0.30		

4 Summary

The results show that the most expressive speakers, those who produced the widest range in acoustic changes, were perceived as more EXTROVERTED and CONSCIENTIOUS. F1, F0 and intensity range, with F0 register – acoustic parameters linked to the speaker's arousal and vocal effort (Scherer 2009; Rilliard et al. 2018) are related to the perception of EXTROVERSION and CONSCIENTIOUS personalities, consistent with hypotheses formulated by Ohala (1994) relating changes in F0 to dominant behavior, and Gussenhoven (2004) relating vocal effort to attention and care.

Ohala's frequency code is related to the F0 register, while Gussenhoven Effort Code relates to F0 span. Speakers with high noise levels in the voice were judged with low AGREEABLENESS, and best produce the expressions involving an imposition on the interlocutor. Correlation of noise (i.e. irregularities in voice) and expressions of impositions are reported in the literature: for example, Goudbeek and Scherer (2010) report spectral noise for the general potency/control dimension and shimmer for low-activation potency (which may to some extend correlate to the expression of SEDU in this study – cf. Rilliard et al. 2018 for another report of noise in charming voices). Finally, speakers judged as having high NEUROTICISM and low OPENNESS were perceived as the best performers for expressions with strong social constraints. The MFA model does not find significant relations between central tendencies and dispersions in acoustic parameters with this third dimension; yet note that these expressions (mostly WOEG, SINC) are highly culturally encoded as can be seen in the long time frame needed for children to acquire them (Shochi et al. 2009a, b) and therefore subject to complex acoustic encodings not reflected in mean and dispersion tendencies (examples of such fine cues could be found e.g. for irony in González-Fuente et al. 2015).

Many questions remain to be answered. For instance, why was seduction grouped with expressions involving an imposition on the interlocutor? What about gender differences? Is there a link with L2 proficiency? Is there an effect of "foreign speaker"? That is, we found a tendency for foreign speakers (French, Japanese) to be rated by U.S. American subjects as less extroverted. This raises the question as to what extent culture affects the perception of personality. More specifically, would French or Japanese listeners judge these personalities the same ways as American listeners? Also, there may be different cultural valences assigned to traits. For instance, in American culture, extroversion and an expansive range of acoustic changes are seen as positive characteristics; however, the contrary may be true in Japan. Extroversion might be perceived more negatively, especially perhaps in professional situations, where being conscientious may be more valued than being extroverted. And, finally, future research is needed to tease out which modality (audio or visual) more readily conveys which of the five traits. Our hope is that this pilot study will stimulate exploration into the topic of how voice (and face) changes accompanying social affective expressions affect judgments of personality.

Acknowledgements. The authors are deeply indebted to the speakers and listeners from Waseda, Bordeaux, New Mexico, and Black Hills State Universities who participated in these experiments. Also, we acknowledge the French ANR) PADE and SEDUCTION CPU (ANR-10-IDEX-03-02) grants, the Japan Society for the Promotion of Science (JSPS), Grants-in-Aid for Scientific Research (A) #25240026 to the first author, and the Brazilian CNPq agency (bolsa PV 406177/2015-5) to the second author. Part of the contents of this paper appeared in Madureira, S. (Ed.) Sonoridades - Sonorities. 1. São Paulo: Edição da Pontifícia Universidade Católica de São Paulo, 2016.

References

Bécue-Bertaut, M., Pagès, J.: Multiple factor analysis and clustering of a mixture of quantitative, categorical and frequency data. Comput. Stat. Data Anal. **52**(6), 3255–3268 (2008)

Borkenau, P., Liebler, A.: The cross-modal consistency of personality: inferring strangers' traits from visual or acoustic information. J. Res. Pers. **26**(2), 183–204 (1992)

Costa, P.T., McCrae, R.R.: Four ways five factors are basic. Personality Individ. Differ. **13**(6), 653–665 (1992)

Degottex, G., Kane, J., Drugman, T., Raitio, T., Scherer, S.: COVAREP - a collaborative voice analysis repository for speech technologies. In: Proceedings of IEEE International Conference on Acoustics, Speech and Signal Processing (ICASSP 2014), pp. 960–964 (2014)

Delattre, P.: Comparing the prosodic features of English, German, Spanish and French. In: IRAL-International Review of Applied Linguistics in Language Teaching, vol. 1(1), 193–210 (1963)

Ekman, P., Friesen, W.V., O'Sullivan, M., Scherer, K.: Relative importance of face, body, and speech in judgments of personality and affect. J. Pers. Soc. Psychol. **38**(2), 270–277 (1980)

González-Fuente, S., Escandell-Vidal, V., Prieto, P.: Gestural codas pave the way to the understanding of verbal irony. J. Pragmat. **90**, 26–47 (2015)

Goudbeek, M., Scherer, K.R.: Beyond arousal: valence and potency/control cues in the vocal expression of emotion. J. Acoust. Soc. America **128**(3), 1322–1336 (2010)

Gussenhoven, C.: The Phonology of Tone and Intonation. Cambridge University Press (2004)

Halberstadt, A.G., Denham, S.A., Dunsmore, J.C.: Affective social competence. Soc. Dev. **10**(1), 79–119 (2001)

Husson, F., Le, S., Pages, J.: Exploratory Multivariate Analysis by Example Using R. Chapman & Hall/CRC, London (2010)

Léon, P.: Précis de phonostylistique: parole et expressivité. Nathan, Paris (1993)

Lippa, R.: The naive perception of masculinity-femininity on the basis of expressive cues. J. Res. Pers. **12**(1), 1–14 (1978)

Madureira, S., de Camargo, Z.A.: Exploring sound symbolism in the investigation of speech expressivity. In: Proceedings of the 3rd ISCA Workshop ExLing 2010, pp. 105–108 (2010)

Nettle, D.: Personality: What Makes You the Way You Are. Oxford University Press, Oxford (2007)

Ohala, J.J.: The frequency code underlies the sound symbolic use of voice pitch. In: Hinton, L., Nichols, J., Ohala, J.J. (eds.) Sound symbolism, pp. 325–347. Cambridge University Press, Cambridge (1994)

Rilliard, A., Erickson, D., Shochi, T., Moraes, J.A.: Social face to face communication - American English attitudinal prosody. In: Proceedings of Interspeech, pp. 1648–1652 (2013)

Rilliard, A., Erickson, D., Shochi, T., de Moraes, J.A.: US English attitudinal prosody performances in L1 and L2 speakers. In: Proceedings 7th International Conference on Speech Prosody (SP 2014), pp. 895–899 (2014)

Rilliard, A., Erickson, D., de Moraes, J., Shochi, T.: On the varying reception of speakers expressivity across gender and cultures, and inference in their personalities. In: Madureira, S. (ed.) Sonoridades – sonorities, pp. 149–163. Edição da Pontifícia Universidade Católica de São Paulo, São Paulo (2016)

Rilliard, A., d'Alessandro, C., Evrard, M.: Paradigmatic variation of vowels in expressive speech: acoustic description and dimensional analysis. J. Acoust. Soc. America **143**(1), 109–122 (2018)

Scherer, K.R.: Judging personality from voice: a cross-cultural approach to an old issue in interpersonal perception. J. Pers. **40**(2), 191–210 (1972)

Scherer, K.R., Wranik, T., Sangsue, J., Tran, V., Scherer, U.: Emotions in everyday life: probability of occurrence, risk factors, appraisal and reaction patterns. Soc. Sci. Inf. **43**(4), 499–570 (2004)

Scherer, K.R.: Emotions are emergent processes: they require a dynamic computational architecture. Philos. Trans. R. Soc. Lond. B Biol. Sci. **364**(1535), 3459–3474 (2009)

Shochi, T., Rilliard, A., Aubergé, V., Erickson, D.: Intercultural perception of English, French and Japanese social affective prosody. In: Hancil, S. (ed.) The Role of Prosody in Affective Speech, pp. 31–59. Linguistic Insights 97, Peter Lang AG, Bern (2009a)

Shochi, T., Erikson, D., Sekiyama, K., Rilliard, A., Aubergé, V.: Japanese children's acquisition of prosodic politeness expressions. In: Proceedings of the Annual Conference of the International Speech Communication Association (INTERSPEECH 2009), 2009, Brighton, UK, pp. 1743–1746 (2009b)

Sonntag, G.P., Portele, T.: PURR—a method for prosody evaluation and investigation. Comput. Speech Lang. **12**(4), 437–451 (1998)

Swerts, M., Krahmer, E.: Audiovisual prosody and feeling of knowing. J. Mem. Lang. **53**(1), 81–94 (2005)

Uldall, E.: Attitudinal meanings conveyed by intonation contours. Lang. Speech **3**(4), 223–234 (1960)

Speech Emotion Recognition Considering Local Dynamic Features

Haotian Guan[1], Zhilei Liu[1(✉)], Longbiao Wang[1],
Jianwu Dang[1,2(✉)], and Ruiguo Yu[1]

[1] Tianjin Key Laboratory of Cognitive Computing and Application,
Tianjin University, Tianjin, China
{htguan, zhileiliu, longbiao_wang,
dangjianwu, rgyu}@tju.edu.cn
[2] Japan Advanced Institute of Science and Technology, Ishikawa, Japan

Abstract. Recently, increasing attention has been directed to the study of the speech emotion recognition, in which global acoustic features of an utterance are mostly used to eliminate the content differences. However, the expression of speech emotion is a dynamic process, which is reflected through dynamic durations, energies, and some other prosodic information when one speaks. In this paper, a novel local dynamic pitch probability distribution feature, which is obtained by drawing the histogram, is proposed to improve the accuracy of speech emotion recognition. Compared with most of the previous works using global features, the proposed method takes advantage of the local dynamic information conveyed by the emotional speech. Several experiments on Berlin Database of Emotional Speech are conducted to verify the effectiveness of the proposed method. The experimental results demonstrate that the local dynamic information obtained with the proposed method is more effective for speech emotion recognition than the traditional global features.

Keywords: Speech emotion recognition · Local dynamic feature
Prosodic feature · Pitch · Segmentation

1 Introduction

As is all known, speech conveys some additional messages beyond the words, such as emotion or identity of the speaker. With the rapid development of human-computer interaction in recent years, there is a growing interest in the emotion recognition from speech. Recognizing the emotion from the speech is beneficial for machines to communicate with the human. However, this is a problem full of challenges because the expression of emotion varies from one person to another [1].

The traditional method of speech emotion recognition is as follows. In most existing approaches, low-level features of each frame in an utterance are extracted firstly. Then, the statistical features such as mean, maximum, and minimum values of these frames are calculated from the whole utterance. However, taking the features of the whole emotional utterance into account is somewhat unreasonable since human's perception of emotional speech is diverse. Considering the arithmetic capability of

© Springer Nature Switzerland AG 2018
Q. Fang et al. (Eds.): ISSP 2017, LNAI 10733, pp. 14–23, 2018.
https://doi.org/10.1007/978-3-030-00126-1_2

computers, some salient features are usually selected to represent the natures of the emotional speech. Therefore, feature selection is critical to explore features which are more effective for the expression of emotion in order to improve the recognition performance. Finally, these selected salient features are fed into a classifier to conduct the speech emotion classification.

In most of the previous works, global acoustic features of an utterance are usually adopted to eliminate the content differences and reduce the number of features [2]. However, emotional information of the speech is usually characterized by its dynamic changes [3]. In other words, the emotion-related components varies with time, rather than being constant in an utterance. Thus, the utilization of global statistical features alone, which takes statistics of the features in a whole utterance, may disregard some local dynamic information of emotion in speech.

To take such local information into consideration, segmentation is simply used to avoid the shortcomings of global features. There have been some studies working on the segmentation of utterances for the classification of speech emotion. In the work of Schuller et al. [4], several segmentation schemes are proposed. The experimental results show that the combination of global and relative time interval features makes a significant improvement. Jeon et al. [5] compare different segment units (3-words, phrases, and time-based segment) and find that using time-based subsentence segment units outperforms others. Zhang et al. [6] use different segment selection approaches based on entropy, mutual information, and correlation coefficients, which yields better performances. Rao et al. [7] report that the performance due to local prosodic features is above that of global ones. All these previous research reveals the effectiveness of segmental features on speech emotion recognition compared with the global utterance features.

Besides, the prosodic features conveying significant emotional information are utilized and analyzed in many previous studies [2, 8]. Pitch, as one of the prosodic features, has been found discriminative across different emotions, to some extent. For example, the average pitch of the speech with anger or happiness emotion is usually higher than which with sadness or fear emotion. In addition, the contour of pitch also differs among the utterances with different emotions [9].

At the aspect of classifiers utilized in previous research, some unsupervised learning methods are commonly used, such as Gaussian Mixture Model (GMM) in [10]. In addition, Support Vector Machine (SVM), which is a kind of supervised learning method, is employed more because of its capability and performance for modeling small-scale data with fewer parameters to be trained. Its target is to find a hyperplane to distinguish the data. Recently, with the development of deep learning methods, Deep Neural Network (DNN), Deep Belief Network (DBN) and some other deep learning methods, which are based on the perception mechanism of the human brain, are also utilized in speech emotion recognition [11, 12]. However, largescale datasets are necessary for the training of such kind of deep learning methods.

In this paper, time-based segmentation approach, which is to divide an utterance according to the time without taking the lexical information into account, is utilized to capture the temporal information of the emotional speech. The utilization of time-based segmentation achieves higher real-time capability, which can improve the audio stream processing performance to certain degree. And a novel pitch probability distribution, which is obtained by drawing the histogram, is proposed as a local dynamic prosodic feature, since pitch plays an important role in the expression of emotion and the histogram can reflect the distribution of the values to a certain degree. Firstly, the pitch histogram and other acoustic features are extracted from each segment of the utterance. After that, an optional processing of principal components analysis (PCA) is adopted for feature selection. Finally, these selected features are fed into an SVM classifier and the predicted class of emotion are obtained. The proposed framework for speech emotion recognition is illustrated in Fig. 1. Several comparative experiments are designed to validate the effectiveness of the proposed method. Based on the comparison of the experimental results, it can concluded that the combination of segmentation and the pitch probability distribution features, which considers the local dynamic information, achieves better results.

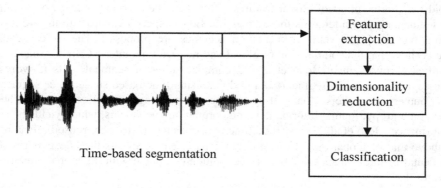

Fig. 1. Overview of the proposed method

The rest of this paper is organized as follows. In Sect. 2, the detailed method is provided, in which the time-based segmentation and the proposed novel pitch probability distribution features obtained by drawing the histogram are introduced. Experimental conditions and results are presented in Sect. 3. Discussion and conclusion are given in Sect. 4.

2 Time-Based Segmentation and Local Dynamic Pitch Probability Distribution Feature Extraction

2.1 Time-Based Segmentation

The Relative Time Intervals (RTI) approach [4] is utilized for time-based speech segmentation. In addition, traditional Global Time Intervals (GTI) approach is adopted for comparison, which simply means using the whole utterance without segmentation and is usually used in traditional methods. Figure 2 shows the illustration of applying GTI and RTI approaches for the utterances with different lengths, in which the strips represent the utterances and the numbers refer to the indexes of the segments.

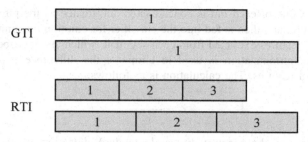

Fig. 2. Global Time Intervals (GTI) and Relative Time Intervals (RTI) approaches applied to a short and a long utterance respectively [4]

In the time-based segmentation approaches, taking the average is one of the simplest techniques to divide an utterance, which also guarantees the same number of the divided segments in an utterance. Therefore, in the RTI approach, as shown in Fig. 3, an utterance is first divided into n segments with the same duration of τ/n, in which τ denotes the length of the utterance, and n keeps invariant in the whole process. Next, each segment will be divided into frames of 25 ms length with 15 ms overlap.

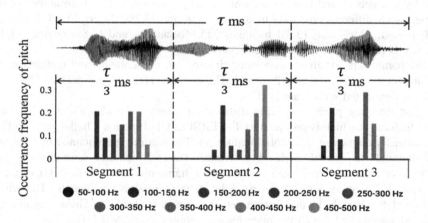

Fig. 3. Local dynamic pitch probability distribution feature extraction $(n = 3)$

2.2 Local Dynamic Pitch Probability Distribution Feature Extraction

After the segmentation, the pitch value of each frame is calculated, and only the values within a certain range are taken into account for pitch histogram computation. As shown in Fig. 3, the horizontal axis corresponds to several bins (or intervals) of the pitch range, while the vertical axis is the occurrence frequency of the pitch falling into each bin. The pitch histogram is normalized, with the sum of the heights equaling one. When the range of pitch is set to $[a, b]$ and the bin width is h, there will be $(b - a)/h$ bins for each segment in the histogram.

Finally, the value of each bin is concatenated and treated as the pitch probability distribution feature, and then is fed into the classifier for emotion recognition, together with some other features extracted from each segment, which are described in the next section. Z-score normalization is used to eliminate the difference in the scales of different kinds of features. The calculation is as follows:

$$x' = \frac{x - \mu}{\sigma} \tag{1}$$

where μ and σ are the mean value and standard deviation of the population, respectively.

3 Experiments

3.1 Experimental Conditions

In this paper, our proposed approach is experimentally evaluated on the commonly used Berlin Database of Emotional Speech (Emo-DB), which contains 535 utterances in German covering seven emotions [13]. Ten sentences without emotional content are acted by five actress and five actors, who are all professional. The distributions of the samples with different emotions in the database are 23.7% anger, 15.1% boredom, 8.6% disgust, 12.9% fear, 13.3% happiness, 11.6% sadness, and 14.8% neutral. 84.3% accuracy is reported for a human perception test.

The parameters for above mentioned time-based segmentation and feature extraction are set as following: $n = 3$, $a = 50\,Hz$, $b = 500\,Hz$, and $h = 50\,Hz$. The pitch values are extracted with Praat [14].

Apart from the pitch probability distribution features, we also use a 384 dimensions' feature set which is provided by INTERSPEECH Emotion Challenge 2009 [15] and is usually employed as a global feature set. The features are obtained by applying 12 functionals to several lowlevel descriptors (LLDs) including zero-crossing rate (ZCR), root mean square (RMS) energy, pitch, harmonics-to-noise ratio (HNR), and MFCC 1–12, together with their first order delta regression coefficients. The whole LLDs and functionals in the feature set are shown in Table 1. These features are extracted automatically with the open resource toolkit openSMILE [16].

Table 1. INTERSPEECH Emotion Challenge 2009 feature set

LLDs (16 × 2)	Functionals (12)
(Δ)ZCR, (Δ)RMS energy, (Δ)pitch, (Δ)HNR, (Δ)MFCC(1–12)	mean, standard deviation, kurtosis, skewness extremes: min/max value, relative min/max, position, range linear regression: offset, slope, MSE

HNR, as one of the LLDs, is computed from the Autocorrelation Coefficient Function (ACF), and can be regarded as voicing probability. It is calculated as:

$$HNR(n) = 10 \log \frac{ACF(T_0)}{ACF(0) - ACF(T_0)} \tag{2}$$

$$ACF(\tau) = \sum_{m=0}^{N-1-\tau} x(m)x(m+\tau) \tag{3}$$

in which N, $x(m)$, and T_0 denote the fundamental period [17], the frame length, and the m th sampling point in the n th frame, respectively.

For the classification model, we used SVM with WEKA 3 Data Mining Toolkit [18]. Linear kernel is applied to avoid overfitting. Leave-one-out cross validation is performed for SVM training and testing to miximize the scale of training data.

In order to evaluate the effectiveness of our proposed method and features, several comparative experiments are conducted from different aspects. Firstly, experiment using global acoustic features of utterances without segmentation is regarded as a benchmark. Then, experiments of different segmentation methods are conducted to verify the effectiveness of our proposed local dynamic pitch probability distribution features. In addition, principal components analysis (PCA) is employed for dimensionality reduction of the features.

3.2 Experimental Results

In this paper, weighted average recall (WA, the number of correctly classified instances divided by the total amount of instances) and unweighted average recall (UA, the mean value of the recall for each class) are used to evaluate the performance of classification, in which the weighted average recall is able to reflect the overall accuracy for imbalanced class.

Explanations:

- No. 1: Commonly-used global features of utterances without segmentation.
- No. 2: Pitch probability distribution features extracted from each segment together with commonly-used global features.
- No. 3: Commonly-used local features extracted from each segment.
- No. 4: Both pitch probability distribution features and commonly-used features extracted from each segment.
- No. 5: Apply PCA with a cumulative contribution rate of 99.0% to the features in No.4.

Table 2 presents the speech emotion recognition accuracies of different comparative experiments. Comparing the results of Experiment 1 without segmentation with the others, we find that time-based segmentation contributes to the accuracy with significant improvements. In addition, the pitch probability distribution features are able to increase the accuracy as well. Furthermore, with segmentation and the pitch probability distribution features applied together, the performance is further improved in

Table 2. Comparative results with different experiments in terms of (un-)weighted average recall (UA/WA)

No.	Feature dimensions	WA (%)	UA (%)
1	384	82.80	81.53
2	384 + 93 = 411	83.55	82.36
3	384 × 3 = 1152	84.11	83.77
4	384 × 3 + 93 = 1179	85.23	84.65
5	1179 → PCA(99.0%) = 409	85.42	84.87

Experiment 4. When the dimensionality is reduced to 409 with the utilization of PCA (cumulative contribution rate: 99.0%), the best result is achieved in Experiment 5, whose relative error rate is 18.08% lower than the benchmark in terms of UA. The improvement is achieved by the local dynamic information extracted with the segmentation approach.

The confusion matrices of the benchmark (Experiment 1) and the best result (Experiment 5) are given in Tables 3 and 4. From the confusion matrices, it can be observed that the performances of our proposed method in Experiment 5 are much better than which in Experiment 1 for most of emotions, which verified the effectiveness of our method.

Table 3. Confusion matrix for Experiment 1 (%)

	Hap.	Neu.	Ang.	Sad.	Fear	Bore.	Dis.
Hap.	**73.24**	1.41	16.9	0	5.63	0	2.82
Neu.	2.53	**84.81**	2.53	0	1.27	8.86	0
Ang.	7.87	0.79	**90.55**	0	0.79	0	0
Sad.	0	1.61	0	**80.65**	0	16.13	1.61
Fear	8.7	1.45	5.8	1.45	**76.81**	1.45	4.35
Bore.	0	2.47	0	9.88	1.23	**86.42**	0
Dis.	4.35	2.17	2.17	0	8.7	4.35	**78.26**

Table 4. Confusion matrix for Experiment 5 (%)

	Hap.	Neu.	Ang.	Sad.	Fear	Bore.	Dis.
Hap.	**73.24**	2.82	16.9	0	7.04	0	0
Neu.	2.53	**91.14**	1.27	0	1.27	3.8	0
Ang.	11.02	0	**88.19**	0	0.79	0	0
Sad.	0	3.23	0	**85.48**	1.61	9.68	0
Fear	7.25	1.45	7.25	0	**81.16**	1.45	1.45
Bore.	0	6.17	0	2.47	1.23	**90.12**	0
Dis.	0	4.35	2.17	4.35	0	4.35	**84.78**

Table 5 gives performances of proposed method in terms of UA on each emotion state. We observe that the segmentation and local dynamic pitch probability distribution features increase the performances of recognition of the majority of emotion states, except for happiness and anger. This result is understandable because happiness and anger utterances have similar dynamic trends on the pitch features [9] and therefore tend to be confused with each other.

Table 5. Effects of proposed method in terms of UA changes on each emotion state (%)

Emotion	Benchmark	Best result	UA changes
Disgust	78.26	84.78	+6.52
Neutral	84.81	91.14	+6.33
Sadness	80.65	85.48	+4.84
Fear	76.81	81.16	+4.35
Boredom	86.42	90.12	+3.70
Happiness	73.24	73.24	0.00
Anger	90.55	88.19	−2.36

Besise, some further experiments are also conducted to explore how the number of segments and interval in the pitch histogram affect the recognition result. The experimental results show that the combination of pitch probability distribution features and commonly-used features extracted from each segment together with dimensionality reduction using PCA (i.e. the experimental program of Experiment 5) achieves the best result in each experimental condition. Table 6 shows the results under the experimental program of Experiment 5 in terms of UA (%). In order to examine the relationship between the number of the segments and the recognition results, experiments with four and five segments for each utterance are conducted, but the results are not as good as that with three segments. In addition, the UA decreases with the increase of the number of the segments. Moreover, the result with a bin width of 50 Hz is better than that of 25 Hz. A possible reason is that with smaller granularity of the segmentation and the pitch probability distribution features extraction, some of the emotional information is counteracted by the content differences in an utterance, and therefore it is adverse to the recognition of speech emotion.

Table 6. Results of different bins in the pitch histogram and number of segments under the experimental program of Experiment 5 in terms of UA (%)

Interval in pitch histogram	3 segments	4 segments	5 segments
50 Hz	84.87	83.16	81.78
25 Hz	82.76	81.85	81.38

4 Discussion and Conclusion

In this paper, a novel local dynamic pitch probability distribution feature is proposed in time-based segments to improve the performance of speech emotion recognition. The experimental results suggest that the local dynamic information obtained by time-based segmentation and pitch probability distribution features are more effective for speech emotion recognition than those traditional global features. Some different segmentation related parameters are also examined in the experiments, the results show that too large or small granularity for the segmentation is adverse to the recognition of speech emotion.

There are several emotional speech corpora in various languages being used in the studies. The common problem, however, is that the scales of them are relatively small with respect to those for Automatic Speech Recognition (ASR), which usually makes it difficult to train the classifier well. Thus, it is also an issue to be addressed that how to achieve ideal performance with small-scaled training data. In addition, pitch is selected as one of the prosodic features that convey important information related to emotion for histogram calculation in this paper. Other features can also be analyzed in the similar way to expect a better performance in our future work.

In this study, we validate the dynamic nature of emotional speech in terms of features. Actually, the classification model influences the performance of recognition in large measure as well. Therefore, in the future, dynamic classification methods such as Recurrent Neural Network (RNN) will be considered since these sequential models are suitable for the dynamic information. Hybrid hierarchical models can also be attempted. Moreover, deep learning methods, which are based on the perception mechanism of the human brain, can be introduced for feature selection instead of traditional PCA method. Also, these features and approaches need to be evaluated on large-scaled dataset in order that the models can be trained enough.

Acknowledgements. The research is supported partially by the National Basic Research Program of China (No. 2013CB329303), and the National Natural Science Foundation of China (No. 61503277 and No. 61303109). The study is supported partially by JSPS KAKENHI Grant (16K00297).

References

1. Johar, S.: Emotional speech recognition. Emotion, Affect and Personality in Speech. SECE, pp. 35–41. Springer, Cham (2016). https://doi.org/10.1007/978-3-319-28047-9_5
2. El Ayadi, M., Kamel, M.S., Karray, F.: Survey on speech emotion recognition: features, classification schemes, and databases. Pattern Recogn. **44**, 572–587 (2011)
3. Wollmer, M., Schuller, B., Eyben, F., Rigoll, G.: Combining long short-term memory and dynamic bayesian networks for incremental emotion-sensitive artificial listening. IEEE J. Sel. Top. Sign. Process. **4**, 867–881 (2010)
4. Schuller, B., Rigoll, G.: Timing levels in segment-based speech emotion recognition. In: INTERSPEECH 2006 - ICSLP, pp. 1818–182. (2006)

5. Jeon, J.H., Xia, R., Liu, Y.: Sentence level emotion recognition based on decisions from subsentence segments. In: IEEE International Conference on Acoustics, Speech and Signal Processing (ICASSP 2011), pp. 4940–4943. IEEE (2011)
6. Zhang, H., Warisawa, S.I., Yamada, I.: An approach for emotion recognition using purely segment-level acoustic features. In: International Conference on Kanesi Engineering and Emotion Research (KEER 2014), pp. 39–49. Linköping University Electronic Press (2014)
7. Rao, K.S., Koolagudi, S.G., Vempada, R.R.: Emotion recognition from speech using global and local prosodic features. Int. J. Speech Technol. **16**, 143–160 (2013)
8. Gangamohan, P., Kadiri, S.R., Yegnanarayana, B.: Analysis of emotional speech—a review. In: Esposito, A., Jain, Lakhmi C. (eds.) Toward Robotic Socially Believable Behaving Systems - Volume I. ISRL, vol. 105, pp. 205–238. Springer, Cham (2016). https://doi.org/10.1007/978-3-319-31056-5_11
9. Ververidis, D., Kotropoulos, C.: Emotional speech recognition: resources, features, and methods. Speech Commun. **48**, 1162–1181 (2006)
10. Ververidis, D., Kotropoulos, C.: Emotional speech classification using Gaussian mixture models. In: IEEE International Symposium on Circuits and Systems (ISCAS 2005), pp. 2871–2874. IEEE (2005)
11. Han, K., Yu, D., Tashev, I.: Speech emotion recognition using deep neural network and extreme learning machine. In: INTERSPEECH 2014. International Speech Communication Association (2014)
12. Sánchez-Gutiérrez, M.E., Albornoz, E.M., Martinez-Licona, F., Rufiner, H.Leonardo, Goddard, J.: Deep learning for emotional speech recognition. In: Martínez-Trinidad, J.F., Carrasco-Ochoa, J.A., Olvera-Lopez, J.A., Salas-Rodríguez, J., Suen, Ching Y. (eds.) MCPR 2014. LNCS, vol. 8495, pp. 311–320. Springer, Cham (2014). https://doi.org/10.1007/978-3-319-07491-7_32
13. Burkhardt, F., Paeschke, A., Rolfes, M., Sendlmeier, W.F., Weiss, B.: A database of german emotional speech. In: INTERSPEECH 2005, pp. 1517–1520. International Speech Communication Association (2005)
14. Boersma, P., Weenink, D.: Praat: doing phonetics by computer (2013)
15. Schuller, B.W., Steidl, S., Batliner, A.: The INTERSPEECH 2009 Emotion Challenge. In: INTERSPEECH 2009, pp. 312–315. ISCA (2009)
16. Eyben, F., Weninger, F., Gross, F., Schuller, B.: Recent developments in openSMILE, the Munich open-source multimedia feature extractor. In: the 21st ACM international conference on Multimedia (MM 2013), pp. 835–838. ACM (2013)
17. Schuller, B.W.: Audio features. Intelligent Audio Analysis, p. 62. Springer, Heidelberg (2013). https://doi.org/10.1007/978-3-642-36806-6_6
18. Hall, M., Frank, E., Holmes, G., Pfahringer, B., Reutemann, P., Witten, I.H.: The WEKA data mining software: an update. ACM SIGKDD Explor. Newslett. **11**, 10–18 (2009)

Commonalities of Glottal Sources and Vocal Tract Shapes Among Speakers in Emotional Speech

Yongwei Li[1], Ken-Ichi Sakakibara[2], Daisuke Morikawa[3], and Masato Akagi[1(✉)]

[1] Japan Advanced Institute of Science and Technology, Ishikawa, Japan
{yongwei,akagi}@jaist.ac.jp
[2] Health Sciences University of Hokkaido, Hokkaido, Japan
kis@hoku-iryo-u.ac.jp
[3] Toyama Prefectural University, Toyama, Japan
dmorikawa@pu-toyama.ac.jp

Abstract. This paper explores the commonalities of the glottal source waves and vocal tract shapes among four speakers in emotional speech (vowel: /a/, neutral, joy, anger, and sadness) based on a source-filter model with the proposed precise estimation scheme. The results are as follows. When compared with the spectral tilts of glottal source waves of neutral, (1) those of anger and joy increased, and those of sadness decreased in the 200- to 700-Hz frequency range; (2) those of anger increased, but those of joy decreased, and those of sadness were the same as those of neutral in the 700- to 2000-Hz range; and (3) all spectral tilts had the same tendency over 2000 Hz. For front vocal tract shapes, the area function of anger was the largest, that of sadness was the smallest, and those of joy and neutral were in the middle.

Keywords: Emotional speech · ARX-LF model · Glottal source waves
Vocal tract shapes

1 Introduction

Emotional-speech recognition and synthesis are topics of major interest in speech-signal processing. Investigating the commonalities among speakers in emotional speech is important for emotional speech-signal processing. Although the commonalities of acoustic features of emotional speech have been investigated and used for emotional speech conversion and recognition [1–3], it was also shown that it is difficult to model emotions by using only these acoustic features [4]. Previous research has suggested that features of speech-production organs, such as glottal source waves and vocal tract shapes, be investigated [4–6]. This requires accurate and independent measurement of glottal source waves and vocal tract shapes. However, the properties of production organs for emotional speech have not been extensively investigated for discussing commonalities. This is because (1) the measurement of vocal-fold vibration and vocal tract shape simultaneously and directly when uttering emotional speech, such as using magnetic resonance imaging (MRI) and electromagnetic articulography

© Springer Nature Switzerland AG 2018
Q. Fang et al. (Eds.): ISSP 2017, LNAI 10733, pp. 24–34, 2018.
https://doi.org/10.1007/978-3-030-00126-1_3

(EMA), is precise but costly [7, 8], and (2) independently and precisely estimating glottal source waves and vocal tract shapes from uttered emotional speech is still challenging.

The aim of this paper is to independently investigate the commonalities of glottal source waves and vocal tract shapes among speakers in emotional speech. To achieve this, three issues are covered: (1) collecting emotional speech data, (2) separating glottal source waves and vocal tract shapes from emotional speech signals, and (3) discussing commonalities among speakers.

2 Database

The vowel sound /a/was uttered by four actors (three males and a female) with different speaking styles including eight emotional states; neutral, happy, sadness, afraid, disgusted, relaxed, surprised, and angry. Neutral was uttered once and the other seven states were uttered three times with different degrees (weak, normal, and strong). Thus, there are a total of 88 utterances ($1 + 7 \times 3 = 22$ for each speaker).

2.1 Data Selection

In order to confirm emotional state of speech from the point of view of the speech perception, a listening experiment was carried out to evaluate the actors' utterances.

There are two types of emotion-evaluation approaches, categorical approach and dimensional approach (valence-arousal [V-A]) [2]. Using the dimensional approach, category and degree of emotion can be described in the V-A space. Since the database stores three different degrees of emotions, the dimensional approach was adopted to evaluate emotions.

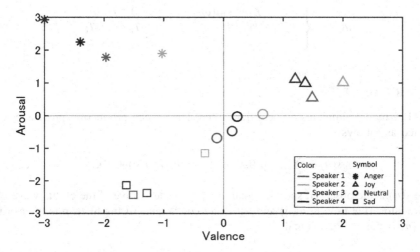

Fig. 1. Selected speech data in V-A space.

Ten people participated in the listening test. For arousal and valence evaluations, a 7-point scale from -3 to 3 (-3: very negative to 3: very positive for valence and -3: very calm to 3: very excited for arousal) was used, and the average evaluation values of the ten participants were calculated. Four basic emotion categories (neutral, joy, sadness, and anger) with strong degrees were selected from the database for further discussion of the commonalities on glottal source waves and vocal tract shapes. The averages of the evaluated values of the selected speech data in the V-A space are shown in Fig. 1.

3   Separation of Glottal Source Waves and Vocal Tract Shapes

To flexibly examine the glottal source waves and vocal tract shapes of emotional speech, a source-filter model was used [9]. Among the source-filter models, the ARX-LF model, which combines the auto-regressive exogenous (ARX) model with the Liljencrant-Fant (LF) model, has been widely used for representing glottal source waves and vocal tract shapes of neutral speech, breathy voice and tense voice [10, 11]. Thus, it is possible to use the ARX-LF model to estimate glottal source waves and vocal tract shapes for emotional speech.

The LF model has six parameters to represent the derivative of the glottal flow: five parameters concerning time T_p, T_e, T_a, T_c, and T_0 and one parameter concerning amplitude E_e, as shown in Fig. 2. The glottal opening instant (GOI) is set to 0, and T_0 is the end of the period, T_p is the phase of the maximum opening of the glottis, T_e is the open phase of the glottis, T_a is the return phase, T_c is end of the return phase, and E_e is the amplitude at the glottal closure instant (GCI) point. The LF model in the time domain is formulated as Eq. 1. Parameters E_1, E_2, a, b and ω are implicitly related to T_p, T_e, T_a, E_e, and T_0 [9].

$$u(t) = \begin{cases} E_1 e^{at} \sin(\omega t) & 0 \le t \le T_e \\ -E_2 \left[e^{-b(t-T_e)} - e^{-b(T_0-T_e)} \right] & T_e \le t \le T_c \\ 0 & T_c \le t \le T_0 \end{cases} \tag{1}$$

3.1   ARX Model

The ARX model simulates a vocal tract filter. The speech production process can be modeled as follows:

$$s(n) + \sum_{i=1}^{p} a_i(n)s(n-i) = b_0(n)u(n) + e(n) \tag{2}$$

where $s(n)$ is the observed speech signal, $u(n)$ is the derivative of the glottal waveform (LF waveform) at time n, $a_i(n)$ and $b_0(n)$ are coefficients of the filter, p is filter order, and $e(n)$ is the residual signal.

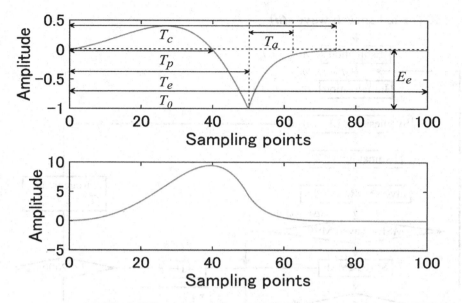

Fig. 2. LF model waveform and glottal waveform

The output signal of the LF model acts as $u(n)$ to the vocal tract filter. Equation (2) is called the ARX model, and the output signal $x(n)$ is a periodic component and $e(n)$ is a non-periodic component in speech.

$$x(n) = -\sum_{i=1}^{p} a_i(n)s(n-i) + b_0(n)u(n) \tag{3}$$

3.2   Scheme of Analysis

The estimation procedure for a period of a glottal source wave is shown in Fig. 3, in which two main processes are included. In the first process, LF parameters and vocal tract coefficients can be obtained with a fixed GCI from a differential electro-glottograph (dEGG) signal as initial values. The initial values of the LF parameters are used for synthesizing $u(n)$. $u(n)$ is then exploited to re-synthesis $x(n)$ using the ARX model with the parameters of the vocal tract filter updated within each period in the mean square error (MSE) sense for $e(n)$ with the help of the least square (LS) method [12]. For the initial values, T_e is estimated from the dEGG signal by searching the GCI and GOI.

In the second process, we want to obtain more accurate LF parameters and vocal tract coefficients. The GCI parameters shift around the initial GCI, and the first process is updated again for the shifted GCI. For the given GCI, the iteration process in the minimization of MSE (MMSE) optimization is set to 2000. The optimal glottal source parameters and vocal tract coefficients are estimated by MMSE.

Recorded speech wave _s(n)_

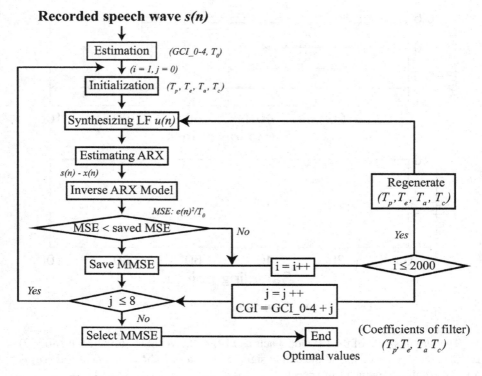

Fig. 3. Estimation scheme of glottal source wave and vocal tract shape

3.3 Evaluation of Proposed Estimation Scheme

A frequently used method for evaluating the estimation algorithm is to estimate synthetic vowels, given the parameter values of synthesized voices. The accuracy of the estimation algorithm can be evaluated by comparing estimated parameter values with referenced values.

Synthesized Data. The speech data used in the evaluation is synthesized, in which derivative glottal waves are synthesized by the LF parameters and the coefficient of vocal tract filter are taken from two vowels /a/and /i/by Kawahara's method [13]. To synthesize speech that can mimic different degrees and types of emotions, the parameters of ARX-LF model (corresponding to the glottal source wave and vocal tract shape) are varied over a wide range. The glottal source parameters are set in a similar way to those of [14, 15], given as $T_e \in [0.3 : 0.05 : 0.9]$, $T_p/T_e \in [0.65 : 0.05 : 0.85]$ and $T_a \in [0.03, 0.08]$. $T_0(F_0)$ is directly taken from real emotional speech, including 18 different F_0, and signal (glottal) to noise ratio (dB) $\in [30 : 10 : 50]$. Thus, the total number of synthesized speech is 14040 periods for /a/ and /i/ $\left(13[T_e] \times 5[T_p] \times 2[T_a] \times 18[F_0] \times 3[\text{SNR}] \times 2[\text{filter}]\right)$.

Results and Discussion. In the analysis, the glottal source waves and vocal tract filters are estimated from the synthesized 14040 periods speech by the ARX-LF model using the proposed scheme.

Examples of the analysis results are shown in Fig. 4, which demonstrates the effectiveness of the analysis approach for decomposing source and filter information. The glottal source wave and vocal tract shape are estimated from one synthesized vowel /a/. The solid lines in Figs. 4(a), (b), and (c) plot the original glottal source wave, original vocal tract shape, and original speech wave, while, the dashed lines plot the estimated glottal source wave, estimated vocal tract shape, and estimated speech wave, respectively. Figures 4(d), (e), and (f) corresponded to (a), (b), and (c) in the frequency domain, respectively. For the vocal tract shape, it was calculated by Wakita's method [16]. A 44100-Hz sampling frequency with a 44th order of the vocal tract filter was applied to synthesize the voice in the synthesis step, while, a 12000-Hz sampling frequency with a 15th order of the vocal tract filter was utilized in the analysis step. Thus, the length of original vocal tract shape (17 cm) is shorter than the length of the estimated vocal tract shape (21 cm) when the assumed sound speed is 340 m/s in the vocal tract (see Fig. 4(b)) when the assumed sound speed is 340 m/s in the vocal tract (see Fig. 4(b)). These results suggest that the estimated glottal source wave, vocal tract shape and speech wave match the original data quite well in both the time and frequency domains.

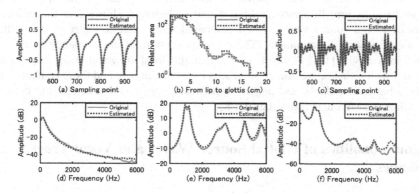

Fig. 4. (a) Original glottal source wave (solid line) and estimated glottal source wave (dashed line), (b) original vocal tract shape (solid line) and estimated vocal tract shape (dashed line), (c) original speech wave (solid line) and estimated speech wave (dashed line); (d) (e) (f) corresponds with (a) (b) (c) in frequency domain, respectively.

The values of (T_p, T_e, T_a, T_c) as glottal source wave parameters and first and second formant frequencies $(F_1$ and $F_2)$ as vocal tract shape parameters were evaluated based on the reference values. The errors (γ) between the reference parameters $\beta \in \{T_p, T_e, T_a, T_c, F_1, F_2\}$ and estimated parameters $\widehat{\beta}$ were calculated as follows:

$$\gamma = \frac{|\widehat{\beta} - \beta|}{\beta} \times 100\% \tag{4}$$

The average errors γ are listed in Table 1, which demonstrates that most of the parameters of the ARX-LF model can be correctly estimated, with errors less than 6% except for those of T_a. Since T_a was the smallest of all parameters and as denominator in Eq. 4, the error of T_a was 34.6%, which indicates the parameters of ARX-LF model can be correctly estimated using the proposed scheme.

Table 1. Average error γ

	T_p	T_e	T_a	T_c	F_1	F_2
/γ/ (%)	5.75	3.41	34.6	5.65	2.05	0.59

3.4 Estimation of Glottal Source Wave and Vocal Tract Shape from Actual Emotional Speech

To discuss the commonalities of glottal source waves and vocal tract shapes among speakers in emotional speech, glottal source waves and vocal tract shapes were estimated from emotional speech uttered by four different speakers (See Fig. 1). The estimated results are illustrated in Figs. 5 and 6. Figure 5 shows that the estimated spectra of glottal source waves, when compared with sadness, anger, and joy, have more high-frequency components. These results show the same tendency as those of the previous report [17]. Figure 6 shows that the front of vocal tract shape, when compared with sadness, area function of anger was the largest, and those of neutral and joy were between sadness and anger. These results are consistent with the previous finds using the MRI and EMA data [7, 8].

4 Commonalities of Glottal Source Waves and Vocal Tract Shapes

Since spectral tilts are frequently used to describe the characteristics of glottal source waves, they were adopted in discussing the commonalities of glottal source wave properties. For vocal tract shapes, the most notable characteristics are the vocal tract area functions. Thus, area functions normalized with the glottis were adopted in discussing the commonalities of vocal tract shape properties.

The commonalities of glottal source waves properties among speakers were summarized from the results shown in Fig. 5. When compared with the spectral tilts of the glottal source waves of neutral, (1) those of anger and joy increased, and those of sadness decreased in the 200- to 700-Hz frequency range; (2) those of anger increased, but those of joy decreased, and those of sadness were the same as those of neutral in the 700- to 2000-Hz range; and (3) all spectral tilts had the same tendency over 2000 Hz.

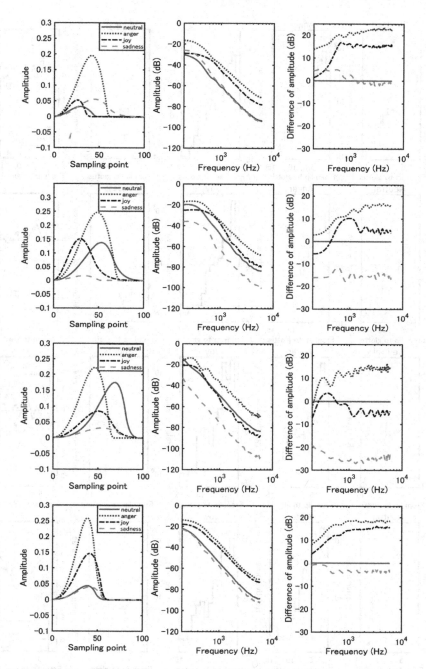

Fig. 5. Results of four speakers (one speaker per row): (a) glottal source waves (first column); (b) spectra of glottal source wave (second column); (c) difference in spectra between neutral and other emotions (third column).

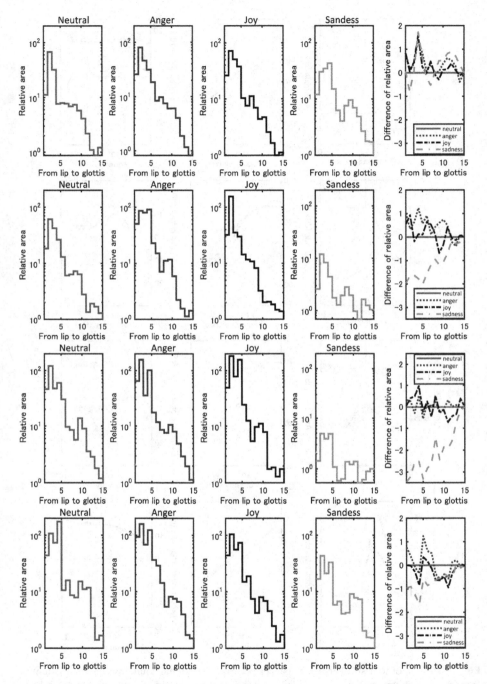

Fig. 6. Results of four speakers (one speaker per row): vocal tract area functions and their differences between neutral and other emotion.

The commonalities of vocal tract shape properties among speakers are summarized in Fig. 6. The width of the front area function of anger was the largest, that of sadness was the smallest, and those of joy and neutral were in the middle.

Moreover, the first formant frequency (F_1) values, which are listed in Table 2, were calculated among speakers and emotions from the ARX model. Table 2 shows that values of F_1 in joy and anger speech are higher than those in neutral and sad speech. Table 2 and Fig. 6 show that a large mouth open area results in a higher F_1 value for anger and joy speech ([8] reported a similar finding using EMA).

Table 2. First formant frequency [Hz] calculated by ARX model

	Anger	Joy	Neutral	Sadness
Speaker 1	861	902	762	650
Speaker 2	920	1020	885	961
Speaker 3	727	791	809	691
Speaker 4	973	861	703	709

5 Conclusion

The commonalities of glottal source waves and vocal tract shapes among speakers in emotional speech were discussed in the following three steps. (1) For collecting emotional speech data, utterances with varying emotions and degrees were extracted from a database and evaluated using the dimensional approach, and four basic emotions (neutral, joy, anger, and sadness) with a strong degree were selected from the V-A emotional space. (2) For separating glottal source waves and vocal tract shapes from emotional speech signals, they were estimated from selected emotional speech data using a proposed precise estimation scheme based on the ARX-LF model. (3) For discussion on the commonalities of glottal source waves and vocal tract shapes among speakers, the spectra of glottal source waves and width of the front area function were investigated. For glottal source waves, spectral tilts of anger and joy increased, and those of sadness decreased in the 200- to 700-Hz frequency range; those of anger increased, but those of joy decreased, and those of sadness were the same as those of neutral in the 700- to 2000-Hz; all spectral tilts had the same tendency over 2000 Hz. For front vocal tract shapes, the area function of anger was the largest, while that of sadness was the smallest, and those of joy and neutral were in the middle. These results are consistent with the previous findings using EMA and MRI.

The results are expected to be used for further discussion on the commonalities of emotional speech glottal source waves and vocal tract shapes among speakers and the applications to emotional speech processing from the point of view of speech production.

The commonalities of four basic emotions with a strong degree were discussed. For future work, we will discuss the commonalities of these emotions with different degrees and attempt to find contributions of glottal source waves and vocal tract shapes to the perception of emotions to further understand emotional speech production mechanisms.

Acknowledgements. This study was supported by a Grant-in-Aid for Scientific Research (A) (No. 25240026) and China Scholarship Council (CSC).

References

1. Schröder, M., Cowie, R., Douglas-Cowie, E., Westerdijk, M., Gielen, S.: Acoustic correlates of emotion dimensions in view of speech synthesis. In: 7th European Conference on Speech Communication and Technology (2001)
2. Hamada, Y., Elbarougy, R., Akagi, M.: A method for emotional speech synthesis based on the position of emotional state in Valence-Activation space. In: Asia-Pacific Signal and Information Processing Association, 2014 Annual Summit and Conference (APSIPA), pp. 1–7. IEEE Press (2014)
3. Li, X., Akagi, M.: Multilingual speech emotion recognition system based on a three-layer model. In: Interspeech, pp. 3608–3612 (2016)
4. Banse, R., Scherer, K.R.: Acoustic profiles in vocal emotion expression. J. Pers. Soc. Psychol. **70**(3), 614–636 (1996)
5. Airas, M., Alku, P.: Emotions in vowel segments of continuous speech: analysis of the glottal flow using the normalised amplitude quotient. Phonetica **63**(1), 26–46 (2006)
6. Gobl, C., Chasaide, A.N.: The role of voice quality in communicating emotion, mood and attitude. Speech Commun. **40**(1–2), 189–212 (2003)
7. Kitamura, T.: Similarity of effects of emotions on the speech organ configuration with and without speaking. In: Interspeech, pp. 909–912, (2010)
8. Erickson, D., Zhu, C., Kawahara, S., Suemitsu, A.: Articulation, acoustics and perception of Mandarin Chinese Emotional Speech. Open Linguist. **2**(1), 620–635 (2016)
9. Fant, G., Liljencrants, J., Lin, Q.-G.: A four-parameter model of glottal flow. in: STL-QPSR 1985, vol. 4, pp. 1–13 (1985)
10. Vincent, D., Rosec, O., Chonavel, T.: Estimation of LF glottal source parameters based on an ARX model. In: Interspeech, pp. 333–336 (2005)
11. Kane, J., Gobl, C.: Evaluation of automatic glottal source analysis. International Conference on Nonlinear Speech Processing, Springer, pp. 1–8 (2013)
12. Ohtsuka, T., Kasuya, H.: Aperiodicity control in ARX-based speech analysis-synthesis method. In: Seventh European Conference on Speech Communication and Technology, pp. 2267–2270 (2001)
13. Kawahara, H., Sakakibara, K.-I., Banno, H., Morise, M., Toda, T., Irino, T.: Aliasing-free implementation of discrete-time glottal source models and their applications to speech synthesis and F0 extractor evaluation. In: Signal and Information Processing Association Annual Summit and Conference (APSIPA), pp. 520–529. IEEE Press (2015)
14. Drugman, T., Bozkurt, B., Dutoit, T.: A comparative study of glottal source estimation techniques. Comput. Speech Lang. **26**(1), 20–34 (2012)
15. Kane, J., Gobl, C.: Evaluation of automatic glottal source analysis. In: Drugman, T., Dutoit, T. (eds.) NOLISP 2013. LNCS (LNAI), vol. 7911, pp. 1–8. Springer, Heidelberg (2013). https://doi.org/10.1007/978-3-642-38847-7_1
16. Wakita, H.: Direct estimation of the vocal tract shape by inverse filtering of acoustic speech waveforms. IEEE Trans. Audio Electroacoust. **21**(5), 417–427 (1973)
17. Schroder M., Cowie R., Douglas-Cowie E., Westerdijk M., Gielen S.C.: Acoustic correlates of emotion dimensions in view of speech synthesis. In: Proceedings of Interspeech 2001, pp. 87–90 (2001)

Articulatory Speech Synthesis

Articulatory Speech Synthesis from Static Context-Aware Articulatory Targets

Anastasiia Tsukanova[1]([⊠]), Benjamin Elie[2], and Yves Laprie[1]

[1] Université de Lorraine, CNRS, Inria, LORIA, 54000 Nancy, France
anastasiia.tsukanova@inria.fr
[2] L2S, CentraleSupelec, CNRS, Université Paris-sud, Université Paris-Saclay,
3 rue Joliot-Curie, 91192 Gif-sur-Yvette, France

Abstract. The aim of this work is to develop an algorithm for controlling the articulators (the jaw, the tongue, the lips, the velum, the larynx and the epiglottis) to produce given speech sounds, syllables and phrases. This control has to take into account coarticulation and be flexible enough to be able to vary strategies for speech production. The data for the algorithm are 97 static MRI images capturing the articulation of French vowels and blocked consonant-vowel syllables. The results of this synthesis are evaluated visually, acoustically and perceptually, and the problems encountered are broken down by their origin: the dataset, its modeling, the algorithm for managing the vocal tract shapes, their translation to the area functions, and the acoustic simulation. We conclude that, among our test examples, the articulatory strategies for vowels and stops are most correct, followed by those of nasals and fricatives. Improving timing strategies with dynamic data is suggested as an avenue for future work.

Keywords: Articulatory synthesis · Coarticulation
Articulatory gestures

1 Introduction

Articulatory speech synthesis is a method of synthesizing speech by managing the vocal tract shape on the level of the speech organs, which is an advantage over the state-of-the-art methods that do not usually incorporate any articulatory information. The vocal tract can be modeled with geometric [2,16,18], biomechanical [1,13] and statistical [9,14] models. The advantage of statistical models is that they use very few parameters, speeding up the computation time. Their disadvantage is that they follow the data a priori without any guidance and do not have access to the knowledge of what is realistic or physically possible. Because of this, to produce correct configurations, they need to be finely tuned.

We were interested in exploring the potential in using quite little, and yet sufficient, static magnetic resonance imaging (MRI) data and implementing one

© Springer Nature Switzerland AG 2018
Q. Fang et al. (Eds.): ISSP 2017, LNAI 10733, pp. 37–47, 2018.
https://doi.org/10.1007/978-3-030-00126-1_4

of the few existing attempts at creating a full-fledged speech synthesizer that would be capable of reproducing the vast diversity of speech sounds.

2   Building an Articulatory Speech Synthesis System

The system is basically made up of three components: the database with the "building blocks" for articulating utterances, the joint control algorithm for the vocal tract and the glottal source, and acoustic simulation. The primary concern of this work are the first two components.

2.1   Dataset

The dataset construction and manipulation were inspired by the work of [3]. We used static MRI data, 97 images collected with a GE Signa 3 T machine with an 8-channel neurovascular coil array. The protocol consisted in a 3D volume acquisition of the vocal tract acquired with a custom modified Enhanced Fast Gradient Echo (EFGRE3D, TR 3.12 ms, TE 1.08 ms, matrix $256 \times 256 \times 76$, with spatial resolution $1.02 \times 1.02 \times 1.0 \ mm^3$). Then we selected the mid-sagittal slice in those images. These data captured articulation without phonation: the speaker was instructed to show the position that he would have to attain to produce a particular sound. For vowels, that is the position when the vowel would be at its clearest if the subject were phonating. For consonant-vowel (CV) syllables, that is the blocked configuration of the vocal tract, as if the subject were about to start pronouncing it. The assumption is that such articulation shows the anticipatory coarticulation effects of the vowel V on the consonant C preceding it. There were 13 vowels, 72 CV syllables and 2 semi-vowels in the final dataset. This covers all main phonemes of the French language, but not in all contexts. Each consonant was recorded in the context of the three cardinal vowels and /y/, which is strongly protruded in French. Some intermediate vocalic contexts were added so as to enable the vowel context expansion algorithm to be checked.

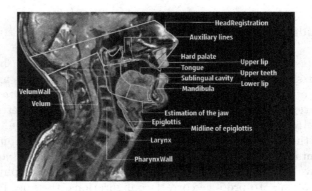

Fig. 1. An example of dataset image annotation (/a/).

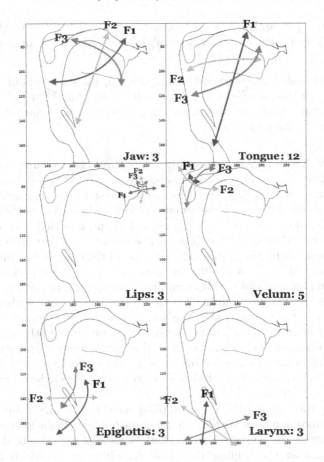

Fig. 2. The PCA-based articulatory model: curve change directions encoded in the first three factors of each articulator (the jaw, the tongue, the lips, the epiglottis, the larynx).

Expanding the Dataset. Since the collected French phonemic dataset was limited, we needed to expand it to cover other contexts as well. We used the notion of the cardinal vowels—/a/, /i/, /u/ and /y/,—assuming that /a/, /i/, /u/ and /y/ represent the most extreme places of vowel articulation, and since then any other vowel articulation can be expanded as a combination of its /a/, /i/ /u/ and /y/ "components". Having captured the C+/a/, C+/i/, C+/u/ and C+/u/context for all consonants C and all non-cardinal vowels V on their own, we were able to estimate the missing C+V samples:

- We projected the vowel V articulatory vector (from $\mathbb{R}^{29}$) onto the convex hull of the /a/, /i/, /u/ and /y/ vectors.
- Assuming that the linear relationship between the C+V vector and the C+/a/, C+/i/, C+/u/ and C+/y/ vectors is the same as the one between V

and /a/, /i/, /u/ and /y/, we estimated C+V from C+/a/, C+/i/, C+/u/ and C+/y/ using the coefficients from the projection of V onto the convex hull of /a/, /i/, /u/ and /y/.

We also estimated the neutral C configuration, the one without any anticipatory effects, as the average of C+/a/, C+/i/, C+/u/ and C+/y/.

Finally, we assumed that the voiced and unvoiced consonants did not have any differences in the articulation.

Articulatory Model. After manually annotating the captures as shown in Fig. 1 we applied a principal-component-analysis (PCA)-based model on the articulator contours [10–12]. We paid special attention to the interaction between articulators and the relevance of deformation modes. Moreover, articulators other than the jaw, tongue and lips are often neglected and modeled with insufficient precision, whereas they can strongly influence acoustics at certain points in the vocal tract. Here are two examples. The position of the epiglottis, which is essentially a cartilage, is likely to modify the geometry of the lower part of the vocal tract by adding an artificial constriction disturbing all the acoustics. It is therefore important to model its deformation modes and interactions with other articulators correctly. In the same way, the velum plays an important role both in controlling the opening of the velopharyngeal port, and in slightly modifying the oral cavity to obtain resonant cavities that give the expected formants of vowels. The acoustic tests we have carried out show in particular that the velum makes it possible to better control the balance between the two cavities necessary for the realization of /u/ and /i/.

Regarding the tongue, PCA was applied on the contours delineated from images. Deformation modes are likely to be impacted by delineation errors. In the case of the tongue, these errors are marginal, or at least give rise to deformation modes coming after the genuine deformations whose amplitude is bigger. On the other hand, the width of epiglottis and/or velum is small on the images, and the errors of delineation, whether manual or automatic, are of the same order of magnitude as genuine deformations. Consequently, PCA applied without precaution will mix both types of deformation. To prevent the apparition of these spurious deformation components the epiglottis was approximated as a thick curve, and only the centerline of epiglottis was analyzed. As a matter of fact, the centerline was determined after delineation of all the epiglottis contours, and the width was set as the average width of all these contours in the upper part where the two epiglottis edges are clearly visible (see Fig. 3). The height of the upper part (where both contours are visible) is adjusted by hand to fit the contours extracted from images. The centerline is approximated as a B-spline and represented by its control points P_l ($0 \leq l < M$ where M is the number of control points) in the form of a two-coordinate vector, and the reconstruction of the epiglottis from the centerline amounts to draw a line at a distance of half the width from the centerline.

The influence of delineation errors is very similar for the velum, which is a fairly fine structure not always well marked on MRI images because it moves

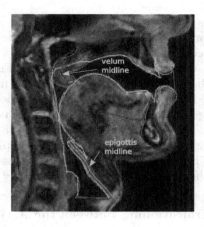

Fig. 3. Epiglottis and velum centerlines reconstructed by the model (solid blue lines). Reconstructed vocal tract is represented by solid red lines. The vocal tract input and output are represented in solid green lines. All these contours are superimposed onto the contours (represented as solid yellow lines) delineated from the image. (Color figure online)

quickly. As for epiglottis we used the centerline and a fairly simple reconstruction algorithm. However, PCA was not applied directly on the control point of the splines because the velum can roll up on itself. This particularity does not lend itself well to the direct use of PCA, which results in the emergence of linear components not appropriate in this case. The centerline is therefore broken down into a series of segments of the same length. Each segment articulates with its predecessor and the first point is fixed. The centerline is then defined as the vector of angles between two consecutive segments. In this way PCA can be applied effectively to velum and gives rise to relevant deformation modes.

The architecture and general organization of the articulatory model are based on the dependency links between the articulators. The main articulator is the jaw which is represented by 3 parameters to get a complete and accurate control. Its geometrical contribution is subtracted from the tongue contours before the application of PCA because tongue is directly attached to the mandible. The tongue is represented by 12 parameters in order to obtain a sufficient precision for the realisation of consonant constrictions. The lips are represented by 3 parameters. Unlike the tongue, the interactions between lips and jaw are more complex. For this reason we subtract the correlation between jaw and lips before applying PCA. The larynx is considered to be independent of the jaw and is represented by 3 parameters to control ist orientation and vertical position. In the same way the velum is considered as an articulator independent of the others. It is analyzed as explained above and is represented by 3 parameters. The epiglottis is the articulator that is subject to the greatest number of influences: the jaw via the tongue, the tongue itself and the larynx. These influences are subtracted by applying a multiple regression to the epiglottis centreline before applying CPA. Analysis of the variance shows that the various influences on the epiglottis account for most of its deformations. Its intrinsic deformation are represented by 3 parameters.

In total these parameters form a vector from $\mathbb{R}^{27}$ (see Fig. 2 for major parameter contributions to the articulator shape). Since the model uses PCA, the zero configuration should correspond to the central position as identified in the dataset, and small changes in the parameter space within a certain neighborhood of zero should correspond to small changes (in terms of distance and shape, not in terms of the resulting acoustics) in the curves. A clipping algorithm is used to solve problems of collision between articulators, i.e. essentially between the tongue and palate. So the model's behavior is not entirely linear.

2.2   Strategies for Transitioning Between the Articulatory Targets

The dataset provided static images capturing idealistic, possibly over-articulated, targets for consonants anticipating particular vowels, whereas the goal was to be also able to deal with consonant clusters and consonants that would not anticipate any vowel at all—for example, due to their ultimate position in a rhythmic phrase. So, in our context, to establish a transitioning strategy would mean three things:

- Choose the building blocks: identify the articulatory target for each phoneme in a phrase. It can either be what was captured in the dataset (a vowel or a consonant assuming vocalic anticipation) or an estimation of what the dataset was missing (missing phonemes, such as voiced consonants, missing contexts or the absence of any context). A consonant cannot anticipate multiple phonemes, nor can vowels anticipate anything due to the restrictions of the dataset at our disposal.
- Decide when — and whether — the articulatory target should be attained.
- Decide how to generate the articulatory positions between the target ones.

Our basic assumption was that by default, consonants anticipate the next coming vowel. However, it would be unrealistic to assume it happens in all cases. This is why we imposed several restrictions on the anticipatory effect:

- Temporal: no coarticulatory effect if the anticipated phoneme is more than 200 ms ahead;
- Spatial: if there is any movement scheduled between the anticipated vowel, the phoneme in question negates the effect. For example, consider such sequence as /lki/: after /l/, the tongue needs to move backward to produce /k/ before coming back forward for /i/. In this situation, our algorithm does not allow the /l/ to anticipate the coming /i/;
- Categorical: it is not possible to anticipate a vowel more than 5 phonemes ahead, and this restriction becomes stricter if it applies across syllable boundaries.

For vowels, there is also a model of target undershoot.

Having established the articulatory targets, the question is how to transition between them. We have tested out three strategies for interpolation between the target vectors:

- Linear: the interpolation between the target vectors is linear;
- Cosine;
- Complex: transitions are done with cosine interpolation, but the timing varies by the articulators. The critical ones reach for their target position faster than the others, while those articulators whose contribution to the resulting sound intelligibility is not as large move slower (for example, the tongue can be in a number of positions for the sound /b/, but the lips have to come into contact). Besides, the articulators composed of heavier tissues (such as the tongue back) move slower than the light and highly mobile ones (such as the lips).

2.3 Obtaining the Sound

Each vocal tract position was encoded in an area function. They were obtained by the algorithm of [7] with coefficients adapted by S. Maeda and Y. Laprie. These parameters only depend on the position in the vocal tract between the glottis and the lips. The transition from the sagittal view to the area function has given rise to several works which contradict each other slightly ([17] and [15]) and it is therefore clear that the determination of the area function will have to take into account the dynamic position of the articulators in the future.

We used an acoustic simulation system implemented by [4] to obtain sound from the area functions and supplementary control files: glottal opening and pitch control.

Glottal opening was modeled by using external lighting and sensing photo-glottography (ePGG) measurements [8]. Within the model, glottis is at its most closed position when producing vowels, nasals and the liquid sound /l/, and momentarily reaches its most open one when producing voiceless fricatives and stops. Voiced fricatives and stops also create peaks in glottal opening, but not as high.

There was no need to model voicing (high-frequency oscillations of low amplitude superimposed onto the glottal opening waves) since the vocal folds operated by the glottal chink model [4,5] are self-oscillating.

3   Evaluation

Each step in the system was evaluated on its own, and afterwards the synthesis results were evaluated visually, acoustically and perceptually. Since the objective of the work was rather to have a fully functional algorithm that produces reasonably realistic movements and sounds rather than to obtain high-quality speech, a more rigorous evaluation, such as a quantitative comparison to the dynamic data on articulatory trajectories, is still an avenue of future work.

3.1   The Articulatory Model and the Trajectories

One peculiarity of the dataset and therefore of the model was the fact that it used only the sagittal section of the speaker's vocal tract. While full three-dimensional models can capture the full geometry of the vocal tract with such phenomena as lateral phonemes (e.g. /l/), two-dimensional models get the benefit of faster computation time and overall simplicity, but irreversibly lose the spatial information.

In general, the articulatory model captured vocal tract positions correctly or with no critical errors, and some adjustments could be necessary only at the points of constriction, since on its own the model did not impose much control over them. This is a disadvantage brought by the nature of the articulatory model: choosing to operate at the level of articulators rather than the resulting vocal tract geometry.

As for the movements, for now we can say that they were reasonable and the coarticulation-affected targets guided the articulators to the positions necessary to produce a particular utterance. One key point here is the timing strategy. Rule-based timing strategy seems to be very rigid for the dynamic nature of speech; it would be more natural to follow speech production processes in humans and to guide the synthesis with the elicited sound or the speaker's expectation—based on their experience—on what this sound will be. We plan to evaluate the transitions with new dynamic MRI data.

3.2   Glottal Opening Control

The algorithm for the glottis opening successfully allowed to distinguish between vowels and consonants. Distinguishing between voiced and voiceless consonants, though, stays a point for improvement, as well as well-coordinated control over the glottis and the vocal tract.

3.3   The Synthesized Sound

Vowels and stops were the most identifiable and correct, although sometimes some minor adjustments in the original data were necessary to obtain formants close to the reference values. When compared to human speech, the formant transitions within the suggested strategies sometimes occurred too fast and sometimes too slowly; again, this highlights the utmost importance of realistic timing strategies. Figure 4 shows an example of the synthesis when it is guided by real timing: /aʃa/ as produced by the system and as uttered by a human. The high-frequency contributions in /ʃ/, not appearing in the human sample, are due to the acoustic simulation. The noise of /ʃ/ is at the correct frequencies, but with a bit different energy distribution, probably because of differences in articulation or in the area functions. There is also an acoustic artifact between /ʃ/ and /a/, which means that more work is necessary on liaising the vocal tract and the source control.

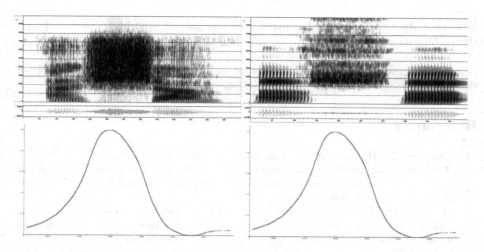

Fig. 4. An example of a human's utterance of /aʃa/ (left) and its synthesis (right) along with the glottal closure control as copied from the EPGG data (below). Phoneme durations are aligned.

4   Conclusion

Regarding speech as a process of transitioning between context-aware targets is an interesting approach that can be connected with the mental processes of speech production: to allow the others to perceive the necessary acoustic cue, the speaker needs to come close enough to the associated articulatory goal. The important difference between a real speaker and the algorithm is the fact that the algorithm solves a static problem, laid out in full; it needs to hit particular targets in a given order. As for humans, we solve a dynamic problem, and coarticulation is not something we put in its definition; rather, coarticulation is our means to make the problem of reaching too many targets in a too short period of time solvable.

The statistically derived articulatory model encodes complicated shapes of the articulators in only 29 parameters, sometimes struggling at the constrictions because of the inherent—and intentional—lack of control over the resulting geometry of the vocal tract.

Those shapes of the articulators change in time according to the produced trajectories of the vocal tract, and those are phonetically sound. Whether there are any important differences between the produced transitions and the ones in real speech, needs to be verified with actual dynamic data.

After the aspect of *how* the articulators move we need to consider *when*. The timing strategies, currently rule-based, apparently need to be extracted from dynamic data, and we can use the approaches by [6] for that.

A closer, intertwined interaction with the acoustic simulation unit—such as guidance on how to navigate between the area functions at the level of separate acoustic tubes and improved control over the glottal opening—could improve the results for consonants.

Acknowledgments. The data collection for this work benefited from the support of the project ArtSpeech of ANR (Agence Nationale de la Recherche), France.

References

1. Anderson, P., Harandi, N.M., Moisik, S., Stavness, I., Fels, S.: A comprehensive 3D biomechanically-driven vocal tract model including inverse dynamics for speech research. In: Sixteenth Annual Conference of the International Speech Communication Association (2015)
2. Birkholz, P., Jackèl, D., Kröger, B.J.: Construction and control of a three-dimensional vocal tract model. In: Proceedings of International Conference on Acoustics, Speech, and Signal Processing (ICASSP 2006), pp. 873–876 (2006)
3. Birkholz, P.: Modeling consonant-vowel coarticulation for articulatory speech synthesis. PloS one **8**(4), e60603 (2013)
4. Elie, B., Laprie, Y.: Extension of the single-matrix formulation of the vocal tract: consideration of bilateral channels and connection of self-oscillating models of the vocal folds with a glottal chink. Speech Commun. **82**, 85–96 (2016)
5. Elie, B., Laprie, Y.: A glottal chink model for the synthesis of voiced fricatives. In: 2016 IEEE International Conference on Acoustics, Speech and Signal Processing (ICASSP), pp. 5240–5244. IEEE (2016)
6. Elie, B., Laprie, Y., Vuissoz, P.A., Odille, F.: High spatiotemporal cineMRI films using compressed sensing for acquiring articulatory data. In: Eusipco, Budapest, pp. 1353–1357, August 2016
7. Heinz, J.M., Stevens, K.N.: On the relations between lateral cineradiographs, area functions and acoustic spectra of speech. In: Proceedings of the 5th International Congress on Acoustics, p. A44 (1965)
8. Honda, K., Maeda, S.: Glottal-opening and airflow pattern during production of voiceless fricatives: a new non-invasive instrumentation. J. Acoust. Soc. Am. **123**(5), 3738–3738 (2008)
9. Howard, I.S., Messum, P.: Modeling the development of pronunciation in infant speech acquisition. Motor Control **15**(1), 85–117 (2011)
10. Laprie, Y., Busset, J.: Construction and evaluation of an articulatory model of the vocal tract. In: 19th European Signal Processing Conference - EUSIPCO-2011. Barcelona, Spain, August 2011
11. Laprie, Y., Vaxelaire, B., Cadot, M.: Geometric articulatory model adapted to the production of consonants. In: 10th International Seminar on Speech Production (ISSP). Köln, Allemagne, May 2014. http://hal.inria.fr/hal-01002125
12. Laprie, Y., Elie, B., Tsukanova, A.: 2D articulatory velum modeling applied to copy synthesis of sentences containing nasal phonemes. In: International Congress of Phonetic Sciences (2015)
13. Lloyd, J.E., Stavness, I., Fels, S.: ArtiSynth: a fast interactive biomechanical modeling toolkit combining multibody and finite element simulation. In: Payan Y. (eds.) Soft Tissue Biomechanical Modeling for Computer Assisted Surgery, pp. 355–394. Springer, Berlin (2012).https://doi.org/10.1007/8415_2012_126

14. Maeda, S.: Compensatory articulation during speech: Evidence from the analysis and synthesis of vocal-tract shapes using an articulatory model. In: Hardcastle, W., Marchal, A. (eds.) Speech Production and Speech Modelling, pp. 131–149. Kluwer Academic Publisher, Amsterdam (1990)
15. McGowan, R., Jackson, M., Berger, M.: Analyses of vocal tract cross-distance to area mapping: an investigation of a set of vowel images. J. Acoust. Soc. Am. **131**(1), 424–434 (2012)
16. Öhman, S.: Coarticulation in VCV utterances: spectrographic measurements. J. Acoust. Soc. Am. **39**(1), 151–168 (1966)
17. Soquet, A., Lecuit, V., Metens, T., Demolin, D.: Mid-sagittal cut to area function tranformations: direct measurements of mid-sagittal distance and area with MRI. Speech Commun. **36**(3–4), 169–180 (2002)
18. Story, B.: Phrase-level speech simulation with an airway modulation model of speech production. Comput. Speech Lang. **27**(4), 989–1010 (2013)

Particle Interaction Adaptivity and Absorbing Boundary Conditions in the Lagrangian Particle Aeroacoustic Model

Futang Wang[1], Qingzhi Hou[1(✉)], Jie Deng[2], Song Wang[3], and Jianwu Dang[1,4]

[1] Tianjin Key Laboratory of Cognitive Computing and Application,
Tianjin University, Tianjin, China
qhou@tju.edu.cn
[2] School of Computer Software, Tianjin University, Tianjin 300354, China
[3] Computational Acoustic Modeling Laboratory,
McGill University, Montreal, Canada
[4] Japan Advanced Institute of Science and Technology, Ishikawa, Japan

Abstract. Recently developed Lagrangian particle aeroacoustic model has shown its capability for simulating acoustic wave propagation problems in flowing fluids. It also has high potential for solving transient acoustics in a domain with moving boundaries. Typical application is sound wave propagation in continuous speech production. When the fluid flow or moving boundary is taken into account, initially evenly distributed particles will become irregular. For irregular particle distribution, the smoothed particle hydrodynamics (SPH) method with constant smoothing length suffers from low accuracy, phase error and instability problems. To tackle these problems, SPH with particle interaction adaptivity might be more efficient, with analog to mesh-based methods with adaptive grids. When the wave arrives at the open boundary, absorbing conditions have also to be applied. Therefore, the main task of this work is to incorporate variable smoothing length and absorbing boundary conditions into the Lagrangian particle aeroacoustic model. The extended model is successfully validated against three typical one- and two-dimensional sound wave propagation problems.

Keywords: Lagrangian particle aeroacoustic model
Continuous speech production · Variable smoothing length
Absorbing boundary

1 Introduction

Acoustic analysis of the vocal tract during human speech has been investigated using many mesh-based methods, such as finite element method (FEM) [1], boundary element method (BEM) [2], and finite-difference time-domain (FDTD) method [3, 4]. They have achieved great success for sound wave propagation in a domain with fixed boundaries. For a domain with moving boundaries (e.g. dynamic vocal tract in continuous speech), however, these mesh-based methods suffer from various difficulties.

© Springer Nature Switzerland AG 2018
Q. Fang et al. (Eds.): ISSP 2017, LNAI 10733, pp. 48–57, 2018.
https://doi.org/10.1007/978-3-030-00126-1_5

Moving mesh can be applied, resulting in annoying disadvantages due to mesh distortion and mesh reconstruction. Immersed boundary method (IBM) is another option as demonstrated by Wei *et al.* [5], but low accuracy and high computational cost are the drawbacks.

Recently, as a promising numerical tool, the meshfree methods are termed as the next generation of numerical methods for solving partial differential equations. As a Lagrangian meshfree method, smoothed particle hydrodynamics (SPH) [6] naturally handles problems with moving boundaries and overcomes the disadvantages due to mesh distortion and reconstruction in moving mesh methods. Zhang *et al.* [7] investigated the application of SPH in acoustic simulation, but they only regarded SPH as a generalized difference method and did not consider particle movement. A Lagrangian aeroacoustic model [8], based on SPH, was developed to solve the problems of acoustic propagation in flowing fluid. However, in this model the situation where the acoustic wave propagates to the open boundary was not considered. In addition, adaptive particle interaction was not included, which can be of high importance when particle distribution becomes highly irregular due to boundary movement. To extend the model's ability to solve more generalized problems, in this paper, we focus on particle interaction adaptivity and absorbing boundary conditions.

2 SPH Formulations of Acoustic Waves

In the Lagrangian framework, the continuity and momentum equations governing sound wave propagation are written as

$$\frac{d\rho}{dt} = -\rho \nabla \cdot \mathbf{v} \tag{1}$$

$$\frac{d\mathbf{v}}{dt} = -\frac{1}{\rho} \nabla p \tag{2}$$

where ρ, $\mathbf{v}$, and p are the density, velocity and sound pressure, respectively. The linear state equation is

$$dp = c_0^2 d\rho \tag{3}$$

where c_0 is the constant sound speed in the medium. With Eqs. (1) and (3), we can get

$$\frac{dp}{dt} = -\rho_0 c_0^2 \nabla \cdot \mathbf{v} \tag{4}$$

where ρ_0 is the reference density of the medium.

The discretized form of the continuity equation for particle a is written as [9]

$$\frac{d\rho_a}{dt} = \sum_{b=1}^{N} m_b \mathbf{v}_{ab} \nabla_a W_{ab} \tag{5}$$

where N is the particle number in the computational domain, m_b is the mass of particle b, $\mathbf{v}_{ab} = \mathbf{v}_a - \mathbf{v}_b$, $W_{ab} = W(\mathbf{x}_a - \mathbf{x}_b, h)$ is the kernel function with smoothing length h and $\nabla_a W_{ab}$ is the kernel gradient with respect to the location of particle a.

Similarly, the momentum equation in SPH form is written as

$$\frac{d\mathbf{v}_a}{dt} = -\sum_{b=1}^{N} m_b \left[\frac{p_a}{\rho_a^2} + \frac{p_b}{\rho_b^2}\right] \nabla_a W_{ab} \tag{6}$$

Note that if Eq. (6) is used to solve problems with shocks, an artificial viscosity term should be added to reduce numerical oscillations [9].

3 Particle Interaction Adaptivity

The basic concept of SPH is to interpolate the quantities of a particle from its neighbors (interpolation points) with a kernel function. The smoothing length h determines the search radius for each particle. In the early development of SPH simulations, the particle interaction distance, i.e., smoothing length, is fixed in space and time. However, the variation of kernel function with respect to the local particle concentration might not be neglected, especially for flows in the fully compressible regime. Monaghan [10] has presented a theoretical description to allow each particle to have its own smoothing length. To better model the aeroacoustic problem with moving boundaries, adaptive particle interaction has to be introduced. Based on the fact that the mass of each particle is constant, i.e., $m_a = \rho_a \Delta x^d = \frac{1}{\xi^d} \rho_a h_a^d$ is fixed, we have

$$h_a = \xi \left(\frac{m_a}{\rho_a}\right)^{\frac{1}{d}} \tag{7}$$

and

$$\frac{dh_a}{dt} = -\frac{1}{d}\frac{h_a}{\rho_a}\frac{d\rho_a}{dt} \tag{8}$$

where Δx is the initial particle spacing, $h = \xi \Delta x$ with a constant ξ, and d is the dimension. Comparing with Eq. (8), the variable smoothing length given by Eq. (7) is simple and does not consider the history information.

The SPH equations for fluids with variable smoothing length have been presented by Nelson and Papaloizou [11] and Monaghan and Rafiee [12]. We simply introduce the main equations used herein.

In SPH the density of particle a is given by a sum over its neighboring particles as

$$\rho_a = \sum_b m_b W_{ab}(h_a) \tag{9}$$

Taking the Lagrangian time derivative of Eq. (9) yields

$$\frac{d\rho_a}{dt} = \frac{1}{\Omega_a} \sum_b m_b (\mathbf{v}_a - \mathbf{v}_b) \nabla_a W_{ab}(h_a) \tag{10}$$

where

$$\Omega_a = 1 + \frac{1}{d} \frac{h_a}{\rho_a} \sum_b m_b \frac{\partial W_{ab}}{\partial h_a} \tag{11}$$

The equation of motion for particle a is given by

$$\frac{d\mathbf{v}_a}{dt} = -\sum_b m_b \left[\frac{P_a}{\Omega_a \rho_a^2} \nabla_a W_{ab}(h_a) + \frac{P_b}{\Omega_b \rho_b^2} \nabla_b W_{ab}(h_b) \right] \tag{12}$$

In this paper, the time stepping is based on the second-order Verlet integrator, of which the details are given by Monaghan and Rafiee [12].

4 Absorbing Boundary

For the simulation of wave propagation in a free space, a computational domain with absorbing boundaries needs to be introduced. Numerical treatment of absorbing boundary conditions has received much attention in many fields, such as electromagnetics [13], shallow water waves [14] and acoustics [15]. Here, we introduce the non-reflecting conditions in the field of acoustics. It is apparent that boundaries will affect the evolution of a specific physical phenomenon that would propagate into the free space. Many methods have been applied to tackle this problem, and the perfectly matched layer (PML) method pioneered by Berenger [16] is mostly used.

Here we use a simple and accurate method recently developed by Modave et al. [17] for simulating linear and non-linear shallow water waves. The basic idea is to add an absorbing layer like a sponge to the surroundings of the domain. In general, the method is implemented by adding a sink term to the governing equations as:

$$\frac{dA}{dt} = f\left(A, \frac{\partial A}{\partial x}, x\right) - \sigma(x)(A - A_{out}) \tag{13}$$

where A is a field variable, $\sigma(A - A_{out})$ is the corresponding sink term, A_{out} is the external boundary value, and $\sigma(x)$ is a damping function depending on the position in the absorbing layer, which differs from zero only in the damping region. Therefore, the continuity Eq. (1) and momentum Eq. (2) become:

$$\frac{d\rho}{dt} = -\rho \nabla \cdot \mathbf{v} - \sigma(x)(\rho - \rho_{out}) \tag{14}$$

$$\frac{d\mathbf{v}}{dt} = -\frac{1}{\rho}\nabla p - \sigma(x)(\mathbf{v} - \mathbf{v}_{out}) \tag{15}$$

Here ρ_{out} and $\mathbf{v}_{out}$ are the density and velocity of the outflow, respectively, and they are both set to be zero herein since in our study the acoustic wave propagates in a medium without mean flow. The damping functions suggested by Molteni et al. [18] might be applied, which decrease the pressure and velocity in the damping layer. Two different choices can be made as

$$\sigma_1(x) = \sigma_0 \frac{L^2 - (x - x_0)^2}{L^2} \tag{16}$$

$$\sigma_2(x) = -\sigma_0 \frac{x - (x_0 + L)}{L} \tag{17}$$

where L is the thickness of the damping layer and x_0 is its starting point. As Modave et al. [17] suggested, we set the coefficient $\sigma_0 = v_{ref}/L$, where v_{ref} is the sound speed. Amongst the two choices, σ_1 is a parabolic function which obtains its maximum at x_0, and σ_2 is a linear function. It can be demonstrated that the above two equations produce zero energy attenuation at the end of the damping layer. In this paper, the parabolic damping function is used for the damping layer and the surroundings of the domain are set with ten absorbing layers. In addition, to ensure that the support of the smoothing kernel is completely covered with particles, we supplement the computational domain with five layers of dummy particles. The pressure of a dummy particle (p_a) is computed using the damping function $p_{ref}\sigma(x)$, where p_{ref} is the end pressure of the damping layer, while the velocity is computed by p_a/c_0.

5   Numerical Results

5.1   Variable Smoothing Length

A Gaussian wave propagating in one-dimensional domain is used to evaluate the performance of the particle interaction adaptivity presented in Sect. 3. The Gaussian sound source is given by

$$p = p_0 e^{-\frac{(x-\mu)}{2\sigma^2}} \tag{18}$$

where $p_0 = 10\,\text{Pa}$, $\mu = 0.1\,\text{m}$ and $\sigma = 0.01$. Other quantities of physical and acoustic model parameters are shown in Table 1.

The SPH solutions with constant and variable smoothing length are shown in Figs. 1a and b, respectively. For comparison, the exact solutions are displayed too. It is clear that with a constant smoothing length, the magnitude of the Gaussian sound wave increases (see Fig. 1a), indicating that the energy of the system is increasing but not conserved. In addition, a slight phase error can be observed. On the other hand, when a variable smoothing length is used, the magnitude of the Gaussian wave remains unchanged, and there is no phase error.

Table 1. Computational parameters.

Parameter	Value
Sound speed	340 m/s
Particle spacing	3.0×10^{-4} m
Smoothing length	4.32×10^{-4} m
Time step	5.0×10^{-7} s
Particle mass	3.0×10^{-4} kg
Simulation time	1.5×10^{-4} s

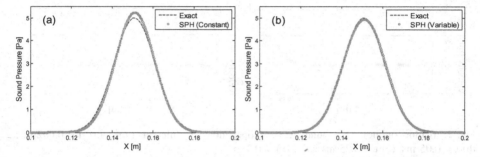

Fig. 1. Comparisons between SPH results with (a) constant and (b) variable smoothing length and theoretical solutions.

5.2 Absorbing Boundary Condition

The one- and two-dimensional Gaussian acoustic waves are used to test the absorbing boundary. In the analog of PML in FDTD, a certain number of absorbing layers are set in the particle model outside of the computational domain. To ensure that the kernel support close to the boundary is fully covered with particles, five layers of dummy particles are supplemented. Numerical results shown in Fig. 2 clearly demonstrate the effectiveness of the implemented PML. The Gaussian wave propagating along the x axis is gradually absorbed (waveform distortion) when it reaches the boundary layers. The numerical results of 2D sound wave propagation with PML absorbing boundaries are shown in Fig. 3. Again, the effectiveness of the damping layer is clearly demonstrated.

To quantitatively evaluate the effectiveness of the absorbing layers, the system energy in the sound field is calculated, which consists of the kinetic energy and potential energy of all particles. The system energy is then written as

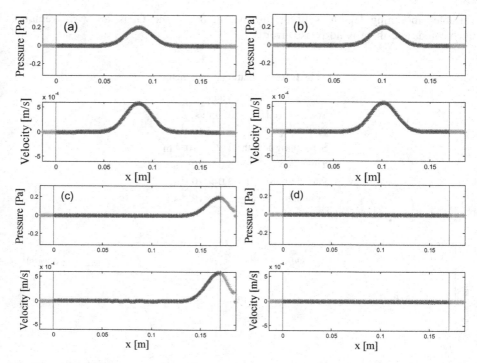

Fig. 2. (Color online) 1D sound wave propagation at different time. (a) t = 0.002 ms, (b) t = 0.05 ms, (c) t = 0.25 ms and (d) t = 0.5 ms.

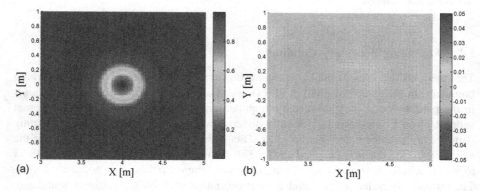

Fig. 3. (Color online) 2D sound wave propagation at different time. (a) t = 0 ms, (b) t = 40 ms.

$$E = E_k + E_p = \sum_i^N \frac{1}{2}\rho_0 V_0 u_i^2 + \frac{1}{2}\frac{p_i^2}{\rho_0 c_0^2} V_0 \tag{19}$$

where E_k is the kinetic energy, E_p is the potential energy, ρ_0 and V_0 are the initial density and volume of the particle, and u and p are the particle velocity and pressure. As the sound pressure in the experiment is too small and the velocity is about 10^{-4}, we use the equation $E_{total} = \sum_i^N p_i^2$ to calculate the energy.

Figure 4 shows the corresponding absorption rate (the ratio between absorbed energy and initial energy) with different layers of PML. It is seen that the absorption rate of the acoustic wave rapidly increases with the number of PML layers and it slows down when the number exceeds seven. Therefore, the PML with seven layers is a reasonable choice, with which up to 99.9% of the energy can be effectively absorbed.

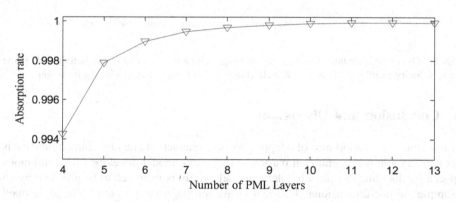

Fig. 4. Absorption rate of different PML layers

5.3   Moving Boundary - Piston Problem

To test the ability of the Lagrangian particle aeroacoustic model to solve moving boundary problems, a moving source (piston) is added to a 1D close-open tube with length of L_t. A sine wave velocity profile is given to the piston at the closed end and zero pressure is given at the open end. The frequencies of harmonic components are compared with the theoretical results given by

$$f = \frac{(2n - 1)c_0}{4L_t} \tag{20}$$

Two different oscillating frequencies, 400 Hz and 600 Hz, around the second harmonic component are tested. The results are shown in Fig. 5. A strong inharmonic frequency component appears corresponding to the oscillating frequency. The closer the oscillation frequency is to the harmonic frequency, the stronger the amplitude of the harmonic components. In addition, an extra experiment is tested where the oscillation frequency is equal to the second harmonic frequency. Due to absence of viscosity and dissipation in the system, the resonance leads the amplitude of the second harmonic component to infinity and the program blows up as expected.

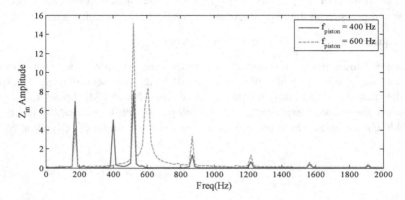

Fig. 5. The input impedances of a close-open pipe with a moving piston oscillating with a sine wave velocity profile at frequency 400 Hz (blue) and 600 Hz (red). (Color figure online)

6   Conclusion and Discussion

In this study, the importance of adaptive particle interaction and absorbing boundary is demonstrated in the Lagrangian particle aeroacoustic model developed for continuous speech production. The particle interaction adaptivity is validated to be effective by an example of one-dimensional Gaussian wave propagation problem. The developed absorbing boundary layer model has good performance in both one- and two-dimensional tests when a reasonable number (seven to ten) of PML layers is used. The ability of the extended particle aeroacoustic model (with variable smoothing length and absorbing boundary) for solving wave propagation problems with moving boundary is demonstrated by a moving piston problem. The extended model has high potential to solve high-dimensional sound propagation problems with moving boundaries, which is the task being worked on.

Acknowledgements. The research is supported partially by the National Basic Research Program of China (No. 2013CB329303), the National Natural Science Foundation of China (No. 51478305 and No. 61233009) and the JSPS KAKENHI Grant (16K00297).

References

1. Aoyama, K., Matsuzaki, H., Miki, N., et al.: Finite-element method analysis of a three-dimensional vocal tract model with branches. J. Acoust. Soc. Am. **100**(4), 2657–2658 (1996)
2. Kagawa, Y., Shimoyama, R., Yamabuchi, T., et al.: Boundary element models of the vocal tract and radiation field and their response characteristics. J. Sound Vib. **157**(3), 385–403 (1992)
3. Takemoto, H., Mokhtari, P., Kitamura, T.: Acoustic analysis of the vocal tract during vowel production by finite-difference time-domain method. J. Acoust. Soc. Am. **128**(6), 3724–3738 (2010)
4. Wang, Y., Wang, H., Wei, J., et al.: Mandarin vowel synthesis based on 2D and 3D vocal tract model by finite-difference time-domain method. In: Proceedings of the 2012 Asia Pacific Signal and Information Processing Association Annual Summit and Conference, Hollywood, CA, pp. 1–4 (2012)
5. Wei, J., Guan, W., Hou, Q., et al.: A new model for acoustic wave propagation and scattering in the vocal tract. In: 17th Annual Conference of the International Speech Communication Association, INTERSPEECH, San Francisco, CA, pp. 3574–3578 (2016)
6. Monaghan, J.J.: An introduction to SPH. Comput. Phys. Commun. **48**(1), 89–96 (1988)
7. Zhang, Y.O., Zhang, T., Ouyang, H., et al.: SPH simulation of sound propagation and interference. In: Proceedings of the 5th International Conference on Computational Method, Cambridge, England, pp. 1–6 (2014)
8. Wei, J., Han, J., Hou, Q., et al.: SPH simulations of aeroacoustic problems in vocal tracts. J. Tsinghua Univ. (Sci. Technol.) **56**(11), 1242–1248 (2016). (in Chinese)
9. Liu, G.R., Liu, M.B.: Smoothed Particle Hydrodynamics: A Meshfree Particle Method. World Scientific, Singapore (2003)
10. Monaghan, J.J.: Smoothed particle hydrodynamics. Ann. Rev. Astron. Astrophys. **30**, 543–573 (1992)
11. Nelson, R.P., Papaloizou, J.C.B.: Variable smoothing lengths and energy conservation in smoothed particle hydrodynamics. Mon. Not. Roy. Astron. Soc. **270**(1), 1–20 (1994)
12. Monaghan, J.J., Rafiee, A.: A simple SPH algorithm for multi-fluid flow with high density ratios. Int. J. Numer. Methods Fluids **71**(5), 537–561 (2013)
13. Berenger, J.P.: A perfectly matched layer for the absorption of electromagnetic waves. J. Comput. Phys. **114**, 185–200 (1994)
14. Vitanza, E., Grammauta, R., Molteni, D., et al.: A shallow water SPH model with PML boundaries. Ocean Eng. **108**, 315–324 (2015)
15. Yuan, X., Borup, D., Wiskin, J.W., et al.: Formulation and validation of Berenger's PML absorbing boundary for the FDTD simulation of acoustic scattering. IEEE Trans. Ultrason. Ferroelectr. Freq. Control **44**(4), 816–822 (2002)
16. Berenger, J.P.: Three-dimensional perfectly matched layer for the absorption of electromagnetic waves. J. Comput. Phys. **127**(2), 363–379 (1996)
17. Modave, A., Deleersnijder, E., Delhez, E.J.M.: On the parameters of absorbing layers for shallow water models. Ocean Dyn. **60**(1), 65–79 (2010)
18. Molteni, D., Grammauta, R., Vitanza, E.: Simple absorbing layer conditions for shallow wave simulations with smoothed particle hydrodynamics. Ocean Eng. **62**(4), 78–90 (2013)

Prediction of F0 Based on Articulatory Features Using DNN

Cenxi Zhao[1], Longbiao Wang[1(✉)], Jianwu Dang[1,2(✉)], and Ruiguo Yu[1]

[1] Tianjin Key Laboratory of Cognitive Computing and Application, Tianjin University, Tianjin, China
{zhaocenxi, longbiao_wang, dangjianwu, rgyu}@tju.edu.cn
[2] Japan Advanced Institute of Science and Technology, Ishikawa, Japan

Abstract. In this paper, articulatory-to-F0 prediction contains two parts, one part is articulatory-to-voiced/unvoiced flag classification and the other one is articulatory-to-F0 mapping for voiced frames. This paper explores several types of articulatory features to confirm the most suitable one for F0 prediction using deep neural networks (DNNs) and long short-term memory (LSTM). Besides, the conventional method for articulatory-to-F0 mapping for voiced frames uses the F0 values after interpolation to train the model. In this paper, only F0 values at voiced frames are adopted for training. Experimental results on the test set on MNGU0 database show: (1) the velocity and acceleration of articulatory movements are quite effective on articulatory-to-F0 prediction; (2) acoustic feature evaluated from articulatory feature with neural networks makes a little better performance than the fusion of it and articulatory feature on articulatory-to-F0 prediction; (3) LSTM models can achieve better effect in articulatory-to-F0 prediction than DNNs; (4) Only-voiced model training method is proved to outperform the conventional method.

Keywords: F0 prediction · Deep neural network · Articulatory features
Recurrent neural network · Long short-term memory

1 Introduction

F0 reflects the significant glottal activities of human speech production systems. Meanwhile, articulatory movements play an important role in generating speech, which involve the systematic motions of a series of apparatus such as tongue, jaw, velum, etc. Researchers have observed the influence of F0 on vowel articulation [1]. Moreover, its known that the velocity and acceleration of the human tongue and lips will become very large when producing stop consonants, especially for unvoiced stops without vocal folds vibration [2]. Thus, there must be some inherent relations between F0 and articulatory feature. F0 prediction based on articulatory features possesses much significance for articulatory controllable speech synthesis [3]. Investigating the relationship between F0 and articulatory feature can also make us a better understanding for how speech is perceived and produced by humans.

© Springer Nature Switzerland AG 2018
Q. Fang et al. (Eds.): ISSP 2017, LNAI 10733, pp. 58–67, 2018.
https://doi.org/10.1007/978-3-030-00126-1_6

In recent years, there has been a good performance using neural networks for predicting F0 from electromagnetic midsagittal articulography (EMA) [4] observations [5]. However, there are two challenging issues to be solved. The first is which input articulatory feature is the most suitable for F0 prediction from non-phonation signals. In paper [5], EMA and the fusion of EMA with mel-generalized cepstrum (MCCs) [6] estimated from EMA (EMA + MCC) are used for F0 prediction. As mentioned above, articulatory velocity and acceleration parameters are also important but they have not been considered for F0 prediction research. The second issue is how to optimize the model training method. In paper [5], F0 interpolation at unvoiced frames is conducted first using an exponential decay function during the articulatory-to-F0 mapping for voiced frames, and then the F0 after interpolation is adopted to train the articulatory-to-F0 for voiced frames mapping model. In fact, we just use F0 at voiced frames when testing the model so only voiced frames make sense in the training process. Considering the interpolated values at unvoiced frames are unbeneficial to this mapping model, there is room to improve the model training method. To solve the two issues, this paper first explores several types of articulatory features to confirm the most suitable one and proposes an only-voiced model training method, in which only voiced frames are used to train the articulatory-to-F0 for voiced frames mapping model.

Articulatory-to-F0 prediction in this paper contains two parts, one part is articulatory-to-voiced/unvoiced flag classification and the other one is articulatory-to-F0 mapping for voiced frames. DNNs and LSTM are adopted for classification and regression in articulatory-to-F0 prediction in this paper. The articulatory features we use include EMA, delta and delta-delta coefficients of EMA, MCC estimated from EMA and the fusion of them. During articulatory-to-F0 mapping for voiced frames, considering only F0 at voiced frames make sense in the training process, we use only-voiced model training method, that is, only voiced frames are adopted to train the model.

The rest of the paper is organized as follows. Section 2 describes a brief overview of the models used in our work, including DNN, RNN and LSTM. Section 3 introduces the proposed articulatory features and an only-voiced model training method in detail. The experimental setup, evaluation results and discussion are given in Sect. 4. Finally, we conclude the research and the future work in Sect. 5.

2 Previous Works

2.1 DNN Based Articulatory-to-F0 Prediction

A DNN is a multi-layer perceptron with several hidden layers between the input layer and the output layer, as illustrated in Fig. 1. DNNs can be trained in a two-stage strategy: the pre-training stage and the fine-tuning stage [7]. For regression tasks, the fine-tuning stage adopts back-propagation (BP) algorithm to minimize the mean square error (MSE) on training set, while for classification tasks, the cost function is set to minimize the cross-entropy (CE).

Consider a sequence of input feature vectors $X = [X_1, X_2, \ldots, X_t]$ and a parallel sequence of output feature vectors $Y = [Y_1, Y_2, \ldots, Y_t]$, where t is the number of frames.

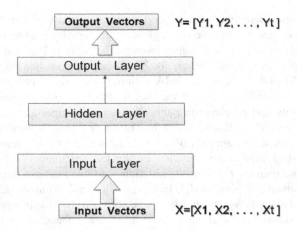

Fig. 1. Illustration of DNN.

In this paper, the input vector X denotes articulatory features. For regression, the output vector Y includes the F0 values for voiced frames and the acoustic feature, 1-st to 40-th orders of MCCs extracted from the waveform using STRAIGHT vocoder [8], for classification, the output vector Y denotes the voiced/unvoiced flags.

2.2 RNNs with LSTM Units Based Articulatory-to-F0 Prediction

RNNs are neural networks with all outputs of the hidden layer feeding back to the input layer, as shown in Fig. 2, where there are cyclical connections among hidden units at the same layer. For fully meshed DNNs, its unable to model the changes in the time sequence. While RNNs can map the history of previous input vectors to each output vector, so that they can make better use of the context information of the input sequence, which makes it more powerful when classifying or generating sequential data. The feedforward processes of RNN are as follows:

$$H_t = \hbar(W_{xh}X_t + W_{hh}H_{t-1} + b_h) \tag{1}$$

$$Y_t = W_{hy}H_t \tag{2}$$

Where $X = [X_1, X_2, \ldots, X_t]$ is the input sequence for time $1, \ldots t$ and the sequences of hidden vectors $H = [H_1, H_2, \ldots, H_t]$, output vectors $Y = [Y_1, Y_2, \ldots, Y_t]$. $\hbar$ is the activation function, W and b denote weight matrices and bias vectors respectively. The RNN can be trained by the back propagation through time (BPTT) algorithm [9].

However, conventional RNNs have the vanishing gradient problem. To solve this problem, long short term memory (LSTM) [10] architecture is proposed. An LSTM unit has three gates, namely input gates, output gates and forget gates, whose functions are to read, write, and reset, as shown in Fig. 3. An LSTM cell can be defined as follows:

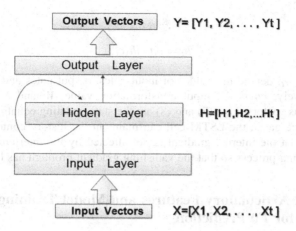

Fig. 2. Illustration of RNN.

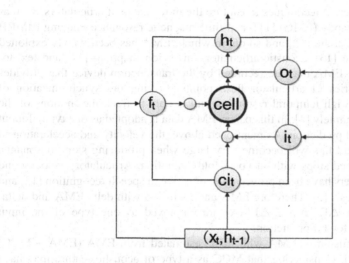

Fig. 3. Illustration of LSTM cell.

$$i_t = g(W_{xi}X_t + W_{hi}h_{t-1} + b_i) \tag{3}$$

$$f_t = g(W_{xf}X_t + W_{hf}h_{t-1} + b_f) \tag{4}$$

$$o_t = (W_{xo}X_t + W_{ho}h_{t-1} + b_o) \tag{5}$$

$$ci_t = tan(W_{xc}X_t + W_{hc}h_{t-1} + b_{ci}) \tag{6}$$

$$c_t = f_t c_{t-1} + i_t ci_t \tag{7}$$

$$h_t = o_t tanh(c_t) \tag{8}$$

Where i_t, f_t, o_t, denote the values of input gates, output gates and forget gates at time t respectively, ci_t is the input transformation value, W and b denote weight matrices and bias vectors, Eqs. (7) and (8) are the state updating equations. Because of the control of the gates, the LSTM cell can maintain the history context information and guarantee that the internal gradient is not affected by the interference of adverse changes in training process so that the vanishing gradient problem has been overcome.

3 Proposed Articulatory Features and Model Training Methods for F0 Prediction

3.1 Articulatory Features for F0 Prediction

There have been different techniques to capture the movements of articulators, such as surface electromyography (EMG) [11], real-time magnetic resonance imaging (rtMRI) data of synchronized audio [12] and so on, in which, EMA has been widely exploited in speech recognition [13], acoustic-articulatory inversion mapping [14] and text-to-speech synthesis [3]. EMA data are recorded by the transduction device that provides data on the trajectories of articulator flesh points carrying out synchronization of speech. It has quite high temporal resolution and is able to track the motions of the main articulators accurately [4]. In this paper, EMA data is adopted as one type of input for articulatory-to-F0 prediction. As mentioned above, the velocity and acceleration of the human tongue and lips will become very large when producing stop consonants, especially for unvoiced stops without vocal folds vibration. Articulatory velocity and acceleration parameters have been proved quite effective in speech recognition [13] and text-to-speech synthesis [3]. Therefore EMA and its fusion with delta EMA and delta-delta EMA (EMA + $\triangle$EMA + $\triangle\triangle$EMA) are employed as one type of the input articulatory features for F0 prediction.

Meanwhile, the fusion of EMA with MCC estimated from EMA (EMA + MCC) are used in paper [5]. Considering that MCC as a type of acoustic feature may have closer relationship than EMA, we apply MCC estimated from EMA as one type of the input articulatory feature to investigate whether EMA there are redundant. EMA and its fusion with delta EMA and delta-delta EMA and MCC (EMA + $\triangle$EMA +$\triangle\triangle$EMA +MCC) are also used as one type of the input articulatory feature.

3.2 Only-Voiced Model Training Method

As we know, F0 value is always zero for silence and unvoiced frames of a sentence [15]. At those frames, the derivatives of F0 would become zero. At the boundaries of a voiced section and an unvoiced section, the derivatives will become infinity. In both cases, problems arise during training the mapping models. In order to solve this problem, paper [16] proposes a continuation algorithm to define the F0 value at unvoiced section with an exponential decay function. Paper [5] employs this method to do F0 interpolation at unvoiced frames, then the F0 after interpolation is adopted to train the articulatory-to-F0 for voiced frames mapping model. However, we just use F0 at voiced frames when testing the model so only F0 at voiced frames make sense in the training process in fact. As the interpolated F0 values at unvoiced frames are not the real values, they are unbeneficial to this mapping model. Therefore in this paper, only F0 values at voiced frames are used to train this mapping model.

4   Experiments

4.1   Experimental Setup

The MNGU0 database [17] is used in our experiments, which consists of 1263 British English utterances from one male native British English speaker with parallel acoustic and EMA recordings. EMA features are captured from 6 sensors located at tongue dorsum, tongue body, tongue lip, jaw, upper lip and lower lip with a sampling frequency of 200 Hz. For each sensor, the coordinates on the front-to-back axis and the bottom-to-top axis (relative to viewing the speakers head from the front) are used, making a total of 12 static EMA features at each frame. The waveforms were in 16 kHz PCM format with 16 bit precision.

1000 utterances are selected from the database to build the training set for our experiments. The validation set and test set contain 20 and 100 utterances respectively. F0 value, voiced/unvoiced flags and MCC are extracted by STRAIGHT analysis.

The mel-cepstral distortion (MCD) [18] in dB between the ground truth and the estimated MCC is adopted as the objective evaluation criterion, which is given by the following equation:

$$Mel - CD[dB] = \frac{10}{ln10} \sqrt{2 \sum_{d=1}^{n} \left(mc_d^{(t)} - mc_d^{(e)} \right)} \tag{9}$$

where $mc_d^{(t)}$ and $mc_d^{(e)}$ denote the d_{th} coefficient of the target and the estimated mel-cepstrum, respectively.

First EMA-to-MCC conversion experiment is conducted with DNN. In this part, different DNN architectures and different sizes of context window are tested using model selection with the testing set. And the results are shown in Tables 1 and 2. Table 1 shows the EMA-to-MCC conversion result when the size of the context window is 15 frames, which means that the previous 7 frames and the subsequent 7 frames along with the current frame are as inputs. And Table 2 shows the EMA-to-MCC conversion result when the DNN architecture is one hidden layer and 1024 hidden nodes for the layer.

From the two tables, it can be found that when the DNN architecture is tuned to be one hidden layer and 1024 hidden nodes for the layer, and the context window is set to 15 frames, we can get the best results, MCD as 0.5301 dB.

Then we utilize EMA, EMA combined with delta EMA and delta-delta EMA, MCC estimated from EMA, EMA combined with the estimated MCC and the fusion of

Table 1. The result of EMA-to-MCC conversion with different DNN architectures

The number of DNN layer	The number of the nodes for each layer	MCD (dB)
1	512	0.5311
1	1024	0.5301*
1	2048	0.5321
2	512, 512	0.5362
2	1024, 1024	0.5336
2	2048, 2048	0.5376

*: the best result of different DNN architectures or different sizes of context window.

Table 2. The result of EMA-to-MCC conversion with different sizes of context window

The size of context window (frames)	MCD (dB)
1	0.5472
3	0.5372
9	0.5314
13	0.5309
15	0.5301*

*: the best result of different DNN architectures or different sizes of context window.

all to do articulatory-to-voiced/unvoiced classification and articulatory-to-F0 mapping for voiced frames using DNN and LSTM.

The structures of the two DNN models for classification and regression are both tuned to be one hidden layer and 512 hidden nodes for the layer. The structures of the two LSTM models for classification and regression are set to be one hidden layer with 512 hidden nodes and one hidden layer with 1024 hidden nodes, respectively. And the context windows are set to be 15 frames for all these models. Besides, we use the interpolation method and the only-voiced method mentioned above when doing articulatory-to-F0 mapping for voiced frames.

4.2   Experimental Results

The results of voiced/unvoiced classification with different input features are shown in Table 3, and accuracy rate is chosen as the evaluation criterion. Table 4 illustrates the results of articulatory-to-F0 mapping for voiced frames with different input features, and shows the comparison of the interpolation method and the only-voiced method. The root-mean-square error (RMSE) are chosen as the evaluation criterion. And the two tables also compare the performance with DNNs and LSTM models for articulatory-to-F0 prediction.

From the two tables first we can see LSTM achieve better performance than DNN for F0 prediction. And delta and delta-delta coefficients of EMA contributes a lot for the result, which demonstrates that articulatory velocity and acceleration parameters are

Table 3. The accuracy rates (%) of voiced/unvoiced classification with different input features

	EMA	EMA + Δ+ΔΔ	EMA + MCC	EMA + Δ+ΔΔ + MCC	MCC
DNN	80.69	81.67	89.22	91.80	91.89
LSTM	81.52	83.33	91.16*	91.95	93.76**

Δ: delta coefficients of EMA.
ΔΔ: delta-delta coefficients of EMA.
*: the result of Liu, Ling, and Dai (2016).
**: the result of proposed input feature and method.

Table 4. The RMSEs of F0 at voiced frames prediction with different methods and input features

	Method	EMA	EMA + Δ+ΔΔ	EMA + MCC
DNN	Interpolation	17.3186	17.1481	17.2264*
	Only-voiced	14.0516	13.8016	13.8081
LSTM	Interpolation	17.2199	17.1225	17.2119
	Only-voiced	14.0471	13.3866	13.0382
		EMA + Δ+ΔΔ + MCC	MCC	
DNN	Interpolation	17.0002	17.2214	
	Only-voiced	13.4163	13.6854	
LSTM	Interpolation	16.8696	17.1347	
	Only-voiced	12.5700**	13.3677	

Δ: delta coefficients of EMA.
ΔΔ: delta-delta coefficients of EMA.
*: the result of Liu, Ling, and Dai (2016).
**: the result of proposed input feature and method.

quite effective for articulatory-to-F0 prediction. And Table 4 also illustrates that only-voiced method has much improvement than interpolation method on articulatory-to-F0 mapping for voiced frames.

Table 3 shows that MCC feature achieves the best performance for articulatory-to-voiced/unvoiced flag classification. From Table 4, we can also find that MCC feature achieves a better performance than EMA + MCC as input for mapping for voiced frames when using DNN. As an acoustic feature, MCC has a closer relationship with F0 than articulatory feature does, and it even degrades the performance when combining the acoustic feature with articulatory feature. However, when the delta and delta-delta coefficients of EMA are fused into, the degration does not seem to be so remarkable.

It can be seen that EMA + ΔEMA +ΔΔEMA +MCC achieve the best performance for articulatory-to-F0 mapping for voiced frames in Table 4. For EMA + Δ EMA +ΔΔEMA +MCC feature, it has the most fusion of these features, acoustic feature, articulatory feature, the articulatory velocity and acceleration parameters, so that it can achieve the best performance for articulatory-to-F0 mapping for voiced frames.

5 Conclusion and Future Work

In summary, for articulatory-to-voiced/unvoiced flag classification and articulatory-to-F0 mapping for voiced frames, MCC feature and EMA + Δ+$\Delta\Delta$ + MCC achieve the best performance respectively. And MCC may have an edge over other features because of its good performance and lower dimensionality.

Although the articulatory feature and method proposed in this paper outperforms the conventional method for F0 prediction, there still much room for improvement. In the future work, first we will try to utilize different machine learning models, such as bidirectional LSTM which can utilize not only the preceding information, but also the subsequent information to improve the performance of articulatory-to-F0 prediction and then apply this work to articulatory controllable speech synthesis.

Acknowledgements. The research is supported partially by the National Basic Research Program of China (No. 2013CB329303), the National Natural Science Foundation of China (No. 61233009 and No. 61771333) and JSPS KAKENHI Grant (16K00297).

References

1. Bauer, D., Kannampuzha, J., Hoole, P., Kröger, B.J.: Gesture duration and articulator velocity in plosive-vowel-transitions. In: Development of Multimodal Interfaces: Active Listening and Synchrony, Second COST 2102 International Training School, pp. 346–353 (2010)
2. Chen, C., Julian, A.: New methods in continuous Mandarin speech recognition. In: European Conference on Speech Communication and Technology (1997)
3. Haykin, S.: Neural Networks: A Comprehensive Foundation, pp. 71–80 (1994)
4. Hess, W., Douglas, O.: Pitch Determination of Speech Signals: Algorithms and Devices by Wolfgang Hess, pp. 219–240. Springer, Heidelberg (1983). https://doi.org/10.1007/978-3-642-81926-1
5. Hinton, Geoffrey E.: A practical guide to training restricted Boltzmann machines. In: Montavon, G., Orr, Geneviève B., Müller, K.-R. (eds.) Neural Networks: Tricks of the Trade. LNCS, vol. 7700, pp. 599–619. Springer, Heidelberg (2012). https://doi.org/10.1007/978-3-642-35289-8_32
6. Hochreiter, S., Jurgen, S.: Long short-term-memory. Neural Comput. **9**(8), 1735–1780 (2014)
7. Honda, K.: Relationship between pitch control and vowel articulation. Haskins Lab. Status Rep. Speech Res. **73**(1), 269–282 (1983)
8. Kawahara, H.: Speech representation and transformation using adaptive interpolation of weighted spectrum: VOCODER revisited. In: IEEE International Conference on Acoustics, Speech, and Signal Processing, vol. 2. pp. 1303–1306 (2002)
9. Koishida, K., Kobayashi, T., Imai, S., Tokuda, K.: Efficient encoding of mel-generalized cepstrum for CELP coders. In: IEEE International Conference on Acoustics, Speech, and Signal Processing, vol. 2, pp. 1355–1358 (1997)
10. Ling, Z.H., Richmond, K., Yamagishi, J., Wang, R.H.: Integrating articulatory features into hmm-based parametric speech synthesis. In: IEEE Transactions on Audio Speech & Language Processing, vol. 17, no. 6, pp. 1171–1185 (2009)

11. Liu, Z.C., Ling, Z.H., Dai, L.R.: Articulatory-to-acoustic conversion with cascaded prediction of spectral and excitation features using neural networks. In: INTERSPEECH, pp. 1502–1506 (2016)
12. Markov, K., Dang, J., Nakamura, S.: Integration of articulatory and spectrum features based on the hybrid HMM/BN modeling framework. Speech Commun. 48(2), 161–175 (2006)
13. Narayanan, S., Erik, B., Prasanta, K.G., Louis, G., Athanasios, K., Yoon, K., Adam, C.: A multimodal real-time MRI articulatory corpus for speech research. In: INTERSPEECH, pp. 837–840 (2011)
14. Richmond, K., Hoole, P., King, S.: Announcing the electromagnetic articulography (day 1) subset of the mngu0 articulatory corpus. In: INTERSPEECH, pp. 1505–1508 (2011)
15. Schönle, P.W., Gräbe, K., Wenig, P., Höhne, J., Schrader, J., Conrad, B.: Electromagnetic articulography: use of alternating magnetic fields for tracking movements of multiple points inside and outside the vocal tract. Brain Lang. 31(1), 26–35 (1987)
16. Schultz, T.W.: Modeling coarticulation in EMG-based continuous speech recognition. Speech Commun. 52(4), 341–353 (2010)
17. Toda, T., Black, A.W., Tokuda, K.: Statistical mapping between articulatory movements and acoustic spectrum using a Gaussian mixture model. Speech Commun. 50(3), 215–227 (2008)
18. Xie, X., Liu, X., Wang, L., Su, R.: generalized variable parameter HMMs based acoustic-to-articulatory inversion. In: INTERSPEECH, pp. 1506–1510 (2015)

A Hybrid Method for Acoustic Analysis of the Vocal Tract During Vowel Production

Futang Wang[1], Qingzhi Hou[1(✉)], Dingyi Pan[2], Jianguo Wei[3], and Jianwu Dang[1,4(✉)]

[1] Tianjin Key Laboratory of Cognitive Computing and Application, Tianjin University, Tianjin, China
qhou@tju.edu.cn, jdang@jaist.ac.jp
[2] Department of Engineering Mechanics, Zhejiang University, Hangzhou 310027, China
[3] School of Computer Software, Tianjin University, Tianjin 300354, China
[4] Japan Advanced Institute of Science and Technology, Ishikawa, Japan

Abstract. A hybrid method for vocal-tract acoustic simulation is proposed to handle the complex and moving geometries during speech production by combining the finite-difference time-domain (FDTD) method and the immersed boundary method (IBM). In this method, two distinct discrete point systems are employed for discretization. The fluid field is discretized by regular Eulerian grid points, while the wall boundary is represented by a series of Lagrangian points. A direct body force is calculated on the Lagrangian points and then interpolated to the neighboring Eulerian points. To validate the proposed hybrid method, a 2D vocal tract model was set by extracting area function from MRI data obtained for the Mandarin vowel /a/. By simulating acoustic wave in this model, the synthesized vowel was analyzed and the obtained formant frequencies were compared to those of real speech sounds. It is found that the mean absolute error of formant frequencies was 8.17% and better than the result in Literature. To show the ability of the hybrid method for solving acoustic problems with moving geometry, a pseudo moving boundary problem was designed and the results agree well with the acoustic theory.

Keywords: Sound propagation · Vocal tract · FDTD · IBM

1 Introduction

Acoustic analysis of the vocal tract by solving the wave equation using numerical methods provides a new tool to study human speech and phonetics. Among the common methods such as finite element method (FEM) [1], boundary element method (BEM) [2], and finite-difference time-domain (FDTD) method [3, 4], FDTD stands out due to its numerical advantages including second-order accuracy, high efficiency and simple orthogonal grids. Using the FDTD method on 2D and 3D vocal tract model, Takemoto et al. [3] performed detailed acoustic analysis of the vocal tract for five Japanese vowels, and Wang et al. [4] for five Mandarin vowels. Although grid points

© Springer Nature Switzerland AG 2018
Q. Fang et al. (Eds.): ISSP 2017, LNAI 10733, pp. 68–77, 2018.
https://doi.org/10.1007/978-3-030-00126-1_7

can be placed exactly on the boundary by carefully choosing the grid size, it is difficult for FDTD to handle complex and moving geometries. Grid transformation at boundary regions can be performed, which however increases the computational time and decreases the numerical accuracy.

Peskin [5] proposed the immersed boundary method (IBM) and applied it to complex geometries and moving boundaries of the heart valve [6]. Its main advantage is easy grid resizing without the necessity for grid points standing exactly on the boundary. This method has been successfully implemented to many applications [7–9]. FDTD combined with IBM has been applied to Mandarin vowels with preliminary success that wave penetration through boundaries was not observed [10]. However, direct application of the non-dimensional Euler equations to real vocal tracts is questionable, and the frequency analysis was not given. In this paper, a hybrid method combining FDTD and IBM is employed to simulate acoustic scattering in the vocal tract during vowel production. The transfer functions obtained by frequency analysis are compared with the measurements.

The rest of the paper is organized as follows. In Sect. 2, the governing equations, numerical methods (FDTD and IBM) and treatment of absorbing boundary are introduced. In Sect. 3, construction of the 2D vocal tract and the details of simulation parameters are described. To show the capability of the proposed method for solving problems with moving boundaries, a hypothetical test case is also designed. In Sect. 4, the numerical results were presented and the acoustic analysis of the synthesized vowels was conducted and compared with the measurements and theory. Finally, conclusions were given in Sect. 5.

2   Mathematical Model

2.1   Governing Equations

Acoustic problems are usually considered as a non-viscous flow and thus the governing equations in a homogeneous lossy acoustic medium [3] are

$$\kappa \frac{\partial p}{\partial t} - \alpha p = -\nabla \cdot \boldsymbol{u} \tag{1}$$

$$\rho \frac{\partial u}{\partial t} - \alpha^* \boldsymbol{u} = -\nabla p \tag{2}$$

where p is pressure, $\boldsymbol{u}$ is velocity, ρ is density, $\kappa = \rho c^2$ represents the compressibility of the medium (c is sound speed), and α and α^* are the attenuation coefficient associated with compressibility and density of the medium, respectively. The coefficient α^* is defined as $\alpha \rho / \kappa$ in the absorbing regions surrounding the computational domain, in which however it is generally set to zero.

2.2 Numerical Methods

FDTD Method. Yee [11] proposed a numerical approach named finite-difference time-domain (FDTD) method to solve the electromagnetic wave propagation problems. It has been widely used in other wave propagation fields like acoustics and seismic waves.

In FDTD, the governing Eqs. (1) and (2) are discretized on a staggered grid with the central difference for spatial derivatives and the leap-frog scheme for temporal derivatives as:

$$p^{n+1/2}(i,j) = p_x^{n+1/2}(i,j) + p_y^{n+1/2}(i,j) \tag{3}$$

$$p_x^{n+1/2}(i,j) = \frac{\kappa/\Delta t - \alpha/2}{\kappa/\Delta t + \alpha/2} p_x^{n-1/2}(i,j)$$
$$- \frac{1}{(\kappa/\Delta t + \alpha/2)/\Delta x} (u_x^n(i+1/2,j) + u_x^n(i-1/2,j)) \tag{4}$$

$$p_y^{n+1/2}(i,j) = \frac{\kappa/\Delta t - \alpha/2}{\kappa/\Delta t - \alpha/2} p_y^{n-1/2}(i,j)$$
$$- \frac{1}{(\kappa/\Delta t + \alpha/2)/\Delta y} (u_y^n(i,j+1/2) - u_y^n(i,j+1/2)) \tag{5}$$

$$u_x^{n+1}(i+1/2,j) = \frac{\rho/\Delta t - \alpha^*/2}{\rho/\Delta t - \alpha/2} u_x^n(i+1/2,j)$$
$$- \frac{1}{(\rho/\Delta t + \alpha^*/2)/\Delta x} (p^{n+1/2}(i+1,j) - p^{n+1/2}(i,j)) \tag{6}$$

$$u_y^{n+1}(i,j+1/2) = \frac{\rho/\Delta t - \alpha^*/2}{\rho/\Delta t - \alpha/2} u_y^n(i,j+1/2)$$
$$- \frac{1}{(\rho/\Delta t + \alpha^*/2)/\Delta y} (p^{n+1/2}(i,j+1) - p^{n+1/2}(i,j)) \tag{7}$$

where $p^{n+1/2}(i,j)$ is the sound pressure on grid point (i,j) at time step of $n + 1/2$ and it is the sum of pressure along x and y directions, i.e., $p_x^{n+1/2}(i,j)$ and $p_y^{n+1/2}(i,j)$, $u_x^n(i+1/2,j)$ and $u_y^n(i,j+1/2)$ represent the velocity components in the x and y direction, and Δx and Δy are the grid size.

Immersed Boundary Method. The immersed boundary method was proposed by Peskin [5] and has been used in many complex flow simulations [7, 8]. To better handle the complex and moving boundary, a general treatment is to replace the boundaries with a series of discrete control points (Lagrangian points), while the flow field is discretized by a Cartesian grid (Eulerian points), as shown in Fig. 1a. For Lagrangian point x_k, the force $f(x_k)$ is calculated over these Lagrangian points and then interpolated to the nearest Eulerian points by a linear interpolation procedure (see Fig. 1b), so that it obtains non-zero force term $F(i,j)$ at grid points near the boundary. After the interpolation, the boundaries of the complex geometry have been successfully handled and then the acoustic equations are solved by a finite difference method.

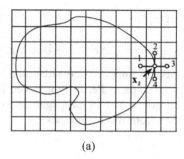
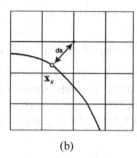

(a)   (b)

Fig. 1. (a) Illustration of Lagrangian force; (b) Distribution of the force to the grid point.

Added Force Field. The Lagrangian force is only added to the momentum equation. All terms are calculated over the Lagrangian points. Since α and α^* are both set to zero in the computational domain, Eqs. (1) and (2) become

$$\kappa \frac{\partial p}{\partial t} + \nabla \cdot \boldsymbol{u} = 0 \tag{8}$$

$$\rho \frac{\partial \boldsymbol{u}}{\partial t} + \nabla p = \mathbf{f}(\mathbf{x}_k) \tag{9}$$

where $\mathbf{f}(\mathbf{x}_k)$ is the Lagrangian force and it can be expressed as

$$\mathbf{f}(\mathbf{x}_k) = \mathbf{f}_a(\mathbf{x}_k) + \mathbf{f}_p(\mathbf{x}_k) \tag{10}$$

in which

$$\mathbf{f}_a(\mathbf{x}_k) = \rho \frac{\partial \boldsymbol{u}}{\partial t} \tag{11}$$

$$\mathbf{f}_p(\mathbf{x}_k) = \nabla p \tag{12}$$

are the acceleration force and pressure force, respectively. For velocity and pressure derivatives in (11) and (12), they are calculated on the Lagrangian points $\mathbf{x}_k$ and the four surrounding auxiliary points (see Fig. 1a). The distances between $\mathbf{x}_k$ and the four auxiliary points are the mesh size (Δx). The velocity and pressure values of these five points are obtained by bi-linear interpolation from the surrounding Eulerian points that enclose the Lagrangian point or virtual points in a grid box [8].

Interpolation. Once the added force on the Lagrangian points is calculated, it needs to interpolate back to the surrounding Eulerian points. For each grid point (i,j) near the boundary, we can identify the closest Lagrangian point $\mathbf{x}_k$. If the distance between the grid point and the related Lagrangian point is greater than the diagonal length of the grid, this point is discarded. Otherwise, the force is interpolated to it by a linear interpolation as:

$$F(i,j) = (1 - ds/h)\mathbf{f}(\mathbf{x}_k) \tag{13}$$

where ds is the distance between these two points and $h = \sqrt{\Delta x^2 + \Delta y^2}$ (see Fig. 1b).

Perfectly Matched Layer. In acoustic simulations, an important consideration is to use a finite domain to properly approximate the infinite space. Perfectly matched layer (PML) [12, 13], which can absorb the radiation sound waves propagating to the boundary, is used herein to minimize the reflections from the boundaries.

3   Experimental Setup

3.1   Constructing 2D Vocal Tract

The study is based on the MRI data obtained for the Chinese mandarin vowel /a/. To obtain the shape of the 3D vocal tract, a series of steps is followed. First, we perform image preprocessing and teeth superimposition on the MRI volumetric images, then convert the images from DICOM to TIFF, denoise using ImageJ software, and finally refer to the teeth superimposition approach proposed by Takemoto [14] to visualize the teeth in MRI images.

The MRI image of the mid-sagittal plane of the vowel /a/with teeth is shown in Fig. 2a, in which two points representing the vocal fold line are manually selected. Then a center line of the vocal tract is automatically calculated from the line [15]. Sections transverse to the line are re-sliced (see Fig. 2b) and the area of vocal tract is measured in each section to obtain the area function.

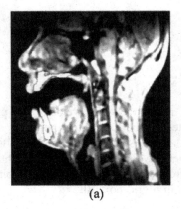 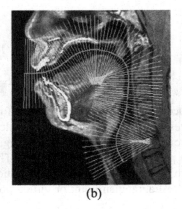

(a)   (b)

Fig. 2.   (a) The MRI image of the mid-sagittal plane of /a/; (b) The locations of cross sections.

Area function of the vocal tract for vowel /a/is given in Fig. 3a. After getting the area function, the model of the 2D vocal tract with area function is constructed to do acoustic simulation experiments. The width of each cross section of the 2D vocal tract is set to be the radius of the corresponding cross section, as shown in Fig. 3b. Then the Cartesian grid for the 2D vocal tract is directly generated.

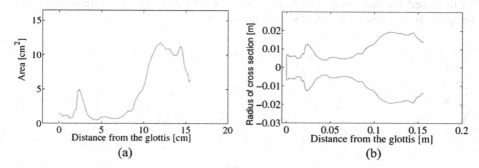

(a)   (b)

Fig. 3. (a) Area function and (b) 2D sketch of the vocal tract for Mandarin vowel /a/.

3.2   Experimental Parameters

In the numerical experiments, eight layers of PML are placed around the computational domain. The simulation parameters are: $\rho = 1.17\,kg/m^3$, $c = 346.3\,m/s$, $\alpha = 100$, $\Delta t = 2 \times 10^{-7}\,s$ and $\Delta x = \Delta y = 1.0\,mm$. The number of Lagrangian points is 329. Note that ρ and c are calculated at temperature $25\,^\circ C$.

A point sound source is set up above the glottis with Gaussian pulse as

$$p(t) = 10\exp\left\{ -\ln2\left[\frac{(x - x_s)^2 + (y - y_s)^2}{b^2} \right] \right\} \tag{14}$$

where (x_s, y_s) is the sound source location and $b = 0.002$. An observation point is set near the center of the lip to record the changing acoustic wave, which is used to calculate the transfer function of vowel. The experiments are carried out on a Windows system, which has 8 central processing units (CPUs) and 24 GB memory.

3.3   Pseudo Boundary in Motion

Compared with grid-based methods, the proposed hybrid method has the potential to handle complex geometries with movement. This arises from the fact that it needs no mesh reconstruction and only requires position changes of the Lagrangian points due to boundary motion. To validate this advantage, a test with pseudo moving boundary has been carried out. In this test, the original area function is manipulated by

$$A_{increased}(x) = [1 + \frac{1}{L}(x - \frac{L}{2})]A_{original}(x) \tag{15}$$

$$A_{descreased}(x) = [1 - \frac{1}{L}(x - \frac{L}{3})]A_{original}(x) \tag{16}$$

for increased and decreased openings (see Fig. 4a), respectively. The corresponding 2D vocal tracts are then obtained as shown in Fig. 4b.

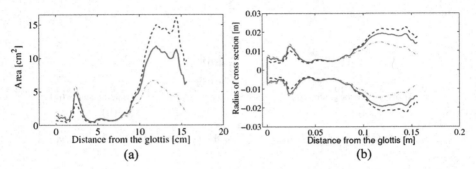

Fig. 4. (Color online) (a) Area function and (b) 2D sketch of the vocal tract for Mandarin vowel /a/: original (solid line), increased opening (dash line) and decreased opening (dash-dot line).

4 Results and Discussion

4.1 Model Validation

Pressure fields at different time levels are shown in Fig. 5. No acoustic wave penetrates the vocal tract as seen in the figure.

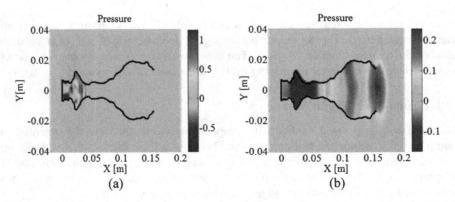

Fig. 5. (Color online) Pressure field at different times of vowel /a/(a) t = 0.1 ms, (b) t = 0.5 ms.

Figure 6 shows the transfer function of the Mandarin vowel /a/. The first four formant frequencies (F1–F4) are 595.5 Hz, 1115 Hz, 2504 Hz and 3470 Hz, respectively. Compared with measurements performed by Wang *et al.* [4] in a sound proof room, the error rates of the first four formant frequencies are −12.5%, 1.2%, −1.3% and −8.7% with a mean absolute error rate of 8.17%. It is 18% smaller than the solution of Wang *et al.* [16], of which the averaged error rate is 10%.

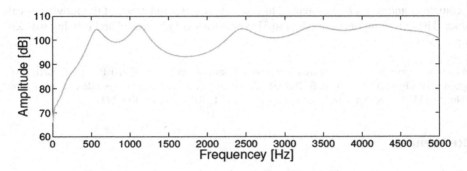

Fig. 6. Transfer function of the Mandarin vowel /a/.

4.2 Pseudo Moving Boundary

Different shapes of vocal tract are simulated with a time change of 20 ms according to the same conditions and the corresponding transfer functions are obtained as shown in Fig. 7. Compared to the first two formants with original vocal tract (F1 = 595.5 Hz and F2 = 1115 Hz), in the increased opening case (dash line), the first formant becomes higher (626 Hz) and the second one becomes slightly lower (1103 Hz); on the other hand, in the decreased opening case (dash-dot line), the first formant becomes lower (511 Hz) and the second becomes higher (1148 Hz). The results meet the basic theory of acoustics.

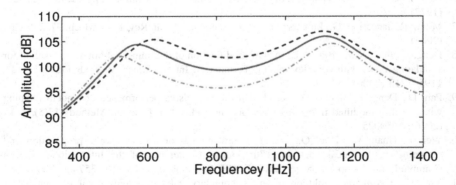

Fig. 7. Comparison of transfer functions of Mandarin vowel /a/ with original and manipulated area functions.

5 Conclusion

In this study, a hybrid method combining FDTD and IBM is developed and used to simulate sound wave propagation in a 2D vocal tract with complex and moving boundaries during the production of Mandarin vowel /a/. The simulated transfer functions are compared with that of the real speech. Numerical results indicate that this hybrid method is an effective way to solve the acoustic problems with complex geometries and it has high potential to calculate transfer functions of the moving vocal tract. The extension of this method to 3D vocal tract is more useful and is being worked on.

Acknowledgements. The research is supported partially by the National Basic Research Program of China (No. 2013CB329303), National Natural Science Foundation of China (No. 61233009 and No. 51478305) and JSPS KAKENHI Grant (16K00297).

References

1. Aoyama, K., Matsuzaki, H., Miki, N., et al.: Finite-element method analysis of a three-dimensional vocal tract model with branches. J. Acoust. Soc. Am. **100**(4), 2657–2658 (1996)
2. Kagawa, Y., Shimoyama, R., Yamabuchi, T., et al.: Boundary element models of the vocal tract and radiation field and their response characteristics. J. Sound Vib. **157**(3), 385–403 (1992)
3. Takemoto, H., Mokhtari, P., Kitamura, T.: Acoustic analysis of the vocal tract during vowel production by finite-difference time-domain method. J. Acoust. Soc. Am. **128**(6), 3724–3738 (2010)
4. Wang, Y., Wang, H., Wei, J., et al.: Mandarin vowel synthesis based on 2D and 3D vocal tract model by finite-difference time-domain method. In: Proceedings of the 2012 Asia Pacific Signal and Information Processing Association Annual Summit and Conference, Hollywood, CA, pp. 1–4 (2012)
5. Peskin, C.S.: Flow patterns around heart valves: a numerical method. J. Comput. Phys. **10**(2), 252–271 (1972)
6. Peskin, C.S.: Numerical analysis of blood flow in the heart. J. Comput. Phys. **25**(3), 220–252 (1977)
7. Mittal, R., Iaccarino, G.: Immersed boundary methods. Annu. Rev. Fluid Mech. **37**, 239–261 (2005)
8. Deng, J., Shao, X.M., Ren, A.L.: A new modification of the immersed-boundary method for simulating flows with complex moving boundaries. Int. J. Numer. Methods Fluids **52**(11), 1195–1213 (2006)
9. Pan, D., Deng, J., Shao, X.M., et al.: On the propulsive performance of tandem flapping wings with a modified immersed boundary method. Int. J. Comput. Methods **13**(5), 1–15 (2016). 1650025
10. Wei, J., Guan, W., Hou, Q., et al.: A new model for acoustic wave propagation and scattering in the vocal tract. In: 17th Annual Conference of the International Speech Communication Association, INTERSPEECH, San Francisco, CA, pp. 3574–3578 (2016)
11. Yee, K.S.: Numerical solution of initial boundary value problems involving maxwell's equations in isotropic media. IEEE Trans. Antennas Propag. **14**(3), 302–307 (1966)

12. Berenger, J.P.: A perfectly matched layer for the absorption of electromagnetic waves. J. Comput. Phys. **114**, 185–200 (1994)
13. Yuan, X., Borup, D., Wiskin, J.W., et al.: Formulation and validation of Berenger's PML absorbing boundary for the FDTD simulation of acoustic scattering. IEEE Trans. Ultrason. Ferroelectr. Freq. Control **44**(4), 816–822 (2002)
14. Takemoto, H., Kitamura, T., Nishimoto, H., et al.: A method of tooth superimposition on MRI data for accurate measurement of vocal tract shape and dimensions. Acoust. Sci. Technol. **25**(6), 468–474 (2004)
15. Takemoto, H., Honda, K., Masaki, S., et al.: Measurement of temporal changes in vocal tract area function from 3D cine-MRI data. J. Acoust. Soc. Am. **119**(2), 1037–1049 (2006)
16. Wang, G., Kitamura, T., Lu, X.: MRI-based study on morphological and acoustic properties of Mandarin sustained vowels. J. Signal Process. **12**, 311–314 (2008)

Considering Lip Geometry in One-Dimensional Tube Models of the Vocal Tract

Peter Birkholz[(✉)] and Elisabeth Venus

Institute of Acoustics and Speech Communication,
TU Dresden, Dresden, Germany
peter.birkholz@tu-dresden.de

Abstract. One-dimensional tube models are an effective representation of the vocal tract for acoustic simulations. However, the conversion of a 3D vocal tract shape into such a 1D tube model raises the question of how to account for the lips, because between the corners of the mouth and the most anterior points of the lips, the cross sections of the vocal tract are open at the sides and hence not well-defined. Here it was examined to what extent simplified tube models of the vocal tract with notches as representations of the lips are acoustically similar to corresponding unnotched models with reduced lengths at the lips end, both with and without teeth. To this end, 3D-printed models of /a, ae, e/ and schwa with different notches and reduced lengths were created. For these, the formant frequencies were measured and analyzed. The results indicate that notched resonators are acoustically most similar to their unnotched counterparts when the length of the unnotched tubes is anteriorly reduced by 50% of the notch depth. However, depending on the formant, vowel, and notch depth, the optimal length reduction can vary between 20–90%.

Keywords: Articulatory speech synthesis · Lip horn · Vocal tract termination

1 Introduction

Despite advanced three-dimensional articulatory models of the vocal tract (e.g., Engwall 2003; Birkholz 2013; Stavness et al. 2012), vocal tract acoustics are mostly simulated in terms of one-dimensional (1D) acoustic tube models that represent the vocal tract as a series of abutting cylindrical tube sections (e.g., Birkholz and Jackèl 2004). The 1D tube models assume plane wave propagation along the vocal tract midline and are much faster to compute than full 3D acoustic simulations. However, the lip region represents a serious difficulty for the 1D approach because the cross-sections of the vocal tract anterior to the corners of the lips are not closed at the sides and hence not defined. This raises the general question of how the "lip horn", i.e., the triangular-shaped space between the corners of the mouth and the most-anterior points of the lips, should be represented in 1D tube models from an acoustic point of view.

From measurements with one subject, Badin et al. (1994) found that the lip horn can be roughly approximated by a single uniform tube section with a length of 11 mm and an area equal to the intra-labial area (in the frontal plane). Lindblom et al. (2007) investigated whether the formant patterns of straight cylindrical tubes with notches at

© Springer Nature Switzerland AG 2018
Q. Fang et al. (Eds.): ISSP 2017, LNAI 10733, pp. 78–86, 2018.
https://doi.org/10.1007/978-3-030-00126-1_8

the "lip end" can be generally reproduced with unnotched tubes of the same diameter and reduced lengths. They found that this was possible when the length of the unnotched tube was the length of the notched tube reduced by about half the notch depth. The optimal length reduction depended somewhat on the tube radius, the notch depth, and the formant index. Lindblom et al. (2010, 2011) furthermore examined physical models with more realistic lip geometries and found that the acoustic effect of the lip horn can be approximated by a length increment that is applied to the last section of the "oral" tube (the tube running from the glottis to the corners of the lips). The length increment is calculated as the "anterior front cavity volume" (the volume anterior to the mouth corners) divided by the cross-sectional area of the most anterior section of the oral tube. However, exactly how the anterior front cavity volume is defined remained unanswered.

In the present study we built upon the investigation by Lindblom et al. (2007), i.e., we examined to what extent it is possible to obtain the formant frequencies of notched tubes with corresponding unnotched tubes of reduced length. We extended the study of Lindblom et al. in two ways:

1. We analyzed not only schwa-like tubes with a constant cross-section, but in addition three two-tube resonators representing the vowels /a, ae, e/.
2. All four resonators were analyzed with and without a row of teeth.

2 Method

2.1 Creation of the Physical Tube Models

In this study we constructed and 3D-printed simplified resonators for the vowels /a, ae, e/ and schwa in nine variants each (one basic variant and eight modified versions). While the basic resonator for schwa was approximated by a single straight cylindrical tube with a constant cross-sectional area of 6 cm^2, /a/, /ae/ and /e/ were approximated by two uniform cylindrical tubes each. The geometries of the basic resonators for /a/ and /ae/ were adopted from Flanagan (1965). The basic geometry of /e/ was based on the /i/ given by Flanagan (1965), but with the cross-sectional area of the anterior tube increased from 1 cm^2 to 2.15 cm^2 to account for the lower tongue position in /e/. We decided to use this /e/ instead of /i/ because the notches for the lips would have unnaturally acute angles for the simplified /i/ resonators. The geometries of all four basic resonators are shown in Fig. 1.

For each vowel, we constructed nine variants as exemplified for the /a/-resonator in Fig. 2. Besides the basic geometry for each vowel, there were four variants where the resonator length was reduced by 1 cm, 2 cm, 3 cm, and 4 cm at the anterior end. The other four variants correspond to the basic resonator with notches of 1 cm, 2 cm, 3 cm, and 4 cm depth at the anterior end. All models contained a small hole of 9 mm diameter at the glottal end to mount a measurement microphone during the acoustic measurements.

Finally, we constructed a simplified row of teeth for each vowel geometry that could be inserted into or removed from the resonators. Based on anatomical data from

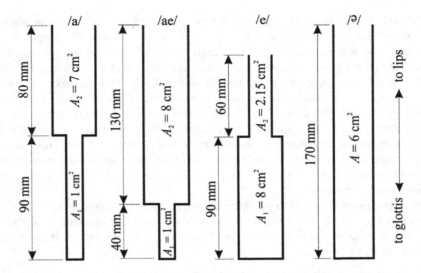

Fig. 1. Dimensions of the four basic vowel resonators.

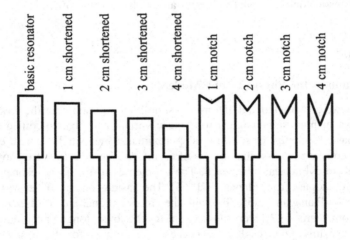

Fig. 2. The nine variants of resonators for /a/.

Slaj et al. (2010), the teeth were modeled as a curved strip where the parabola $y = 0.043 \cdot x^2 - 2.981$ defined the palatal boundary of the teeth row in the axial plane with x and y being the coordinates on the left-right and anterior-posterior axes in mm, respectively, and where $(x, y) = (0, 0)$ is the contact point between the left and right central incisors (see Fig. 3). The teeth strips were 2 mm thick and 1 cm high at the incisors, and inserted at a distance of 2 mm from the most anterior points of the tubes. Figure 3 shows the model variant with the 4 cm notch (maximally spread lips) and inserted teeth, for each vowel.

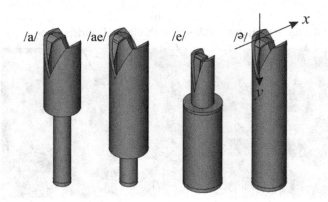

Fig. 3. Resonators with 4 cm notch depth and inserted teeth. The x and y axes define the coordinate system in which the shape of the row of teeth was defined.

All $4 \cdot 9 = 36$ resonators and the teeth rows were manufactured with a 3D-printer (type Ultimaker 2) using the material polylactide (PLA), i.e., the resonators had hard walls. The schwa-resonators were printed in one piece each, and the resonators for /a/, /ae/, and /e/ were printed in two pieces each ("big tube" and "little tube") and then glued together. The wall thickness of all models was 3 mm, and the walls were 100% filled with PLA (100% infill). In previous experiments we found that less than 100% infill made the model walls less stiff, which may cause undesirable resonances in the acoustic transfer functions due to vibrations of the walls.

2.2   Measurement of Formant Frequencies

For each resonator, we measured the volume velocity transfer functions between the glottis and the lips both with and without teeth. The transfer functions were measured with the recently proposed method by Fleischer et al. (2018) that avoids the difficulties of constructing a volume velocity source for glottal excitation. It extends an idea of Kitamura et al. (2009) that is based on the principle of acoustic reciprocity. The method excites the resonances in a given model with a loudspeaker (VISATON speaker, type FR 10-8 Ω, cone diameter 10 cm) about 30 cm in front of the model that emits a broadband sine sweep with a power band of 50–10.000 Hz into the lip opening of the model. At the same time, the sound pressure $P_g(\omega)$ is measured at the glottal end inside the model using a 1/4" measurement microphone (type MK301E/MV310, www. microtechgefell.de) inserted through a hole at the (otherwise closed) glottal end. After this, the lip opening of the resonator is closed with modeling clay, and a second measurement of pressure $P_m(\omega)$ with the same sweep excitation is performed with a microphone centered around 3 mm in front of the closed lips. According to Fleischer et al. (2018), the ratio $P_g(\omega)/P_m(\omega)$ is exactly the volume velocity transfer function between the glottis and the lips, that is typically used to characterize the acoustics of vowels. The spectral resolution of the transfer functions was 1 Hz, and the first four formant frequencies were determined as the peaks of the magnitude spectrum.

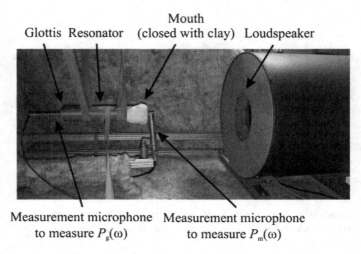

Fig. 4. Setup for the measurement of the acoustic transfer functions, here with a schwa resonator with the lip opening closed with clay to measure $P_m(\omega)$.

The measurement setup is shown in Fig. 4, where the lips have been closed for the measurement of $P_m(\omega)$. All measurements were performed in an anechoic chamber.

Figure 5 shows three of the measured transfer functions for /ae/-resonators with 0 cm, 2 cm, and 4 cm deep notches (without teeth). Apparently, the formant frequencies increase monotonically as the notch depth is increased from 0 cm to 4 cm. Hence, the effect of increasing the notch depth in a tube is similar to the effect that we expect when the length of the tube is reduced (without making a notch).

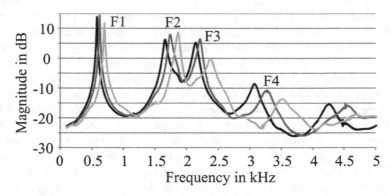

Fig. 5. Measured transfer functions for /ae/ without teeth with 0 cm (black), 2 cm (gray), and 4 cm (light gray) deep notches.

2.3   Calculation of Reduction Factors

For each of the first four formants of the 16 individual notched resonators (4 vowels × 4 notch depths) we determined, based on the acoustic data, what the length reduction of the corresponding basic (unnotched) resonator would need to be in order to produce the same formant frequency. This procedure is illustrated in Fig. 6, which shows the formant frequencies of the /a/-resonators for all four shortening lengths and notch depths, i.e., for all nine tube variants for /a/. The Figure shows that the formant frequencies of the notched resonators (solid lines) are consistently lower than the frequencies of the unnotched resonators with a shortening length equal to the notch depth. For any of the first four formants and any of the four notch depths (1 cm, 2 cm, 3 cm, 4 cm) of the notched resonators, there was one shortening length Δl of the corresponding unnotched resonator that produced the same formant frequency. The gray arrows in Fig. 6 illustrate the calculation of Δl for F_4 of the /a/-resonator with a 3 cm notch. Here, the corresponding shortening length is $\Delta l \approx 1.5$ cm (formant frequencies for shortening lengths between the discrete values of 0 cm, 1 cm, 2 cm, 3 cm, and 4 cm were linearly interpolated). In this way, we calculated a "reduction factor" $k = \Delta l/d$ for each vowel, notch depth d, and formant, both with and without inserted teeth, i.e., $4 \times 4 \times 4 \times 2 = 128$ factors in total. For the example in Fig. 6, the reduction factor is $k \approx 1.5$ cm/3 cm = 0.5. Hence, the notch depth of a notched

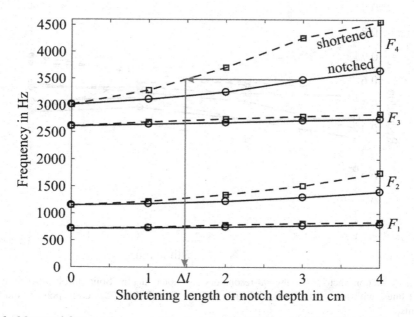

Fig. 6. Measured formant frequencies for the vowel /a/ (without teeth) with the different notch depths (solid lines) and shortened lengths (dashed lines). The gray arrows indicate as an example that for F_4, the resonator with a 3 cm notch is equivalent to the corresponding unnotched resonator shortened by about 1.5 cm.

resonator multiplied by the reduction factor yields the length by which the original unnotched resonator has to be shortened to produce the same formant frequency.

3 Results

Figure 7 illustrates how the reduction factor varies as a function of the notch depth for the vowel /a/. Here, the average reduction factor is 0.55 and varies between 0.23 to 0.75 for a notch depth of 1 cm, and between 0.4 and 0.7 for a notch depth of 4 cm. Furthermore, the overall variation of the reduction factor is greater for the resonators with teeth than for the resonators without teeth. Similar pictures were obtained for the vowels /e/, /ae/, and schwa.

The same trends as shown in Fig. 7 can be observed when the 128 calculated reduction factors are sorted by formants and vowels as in Fig. 8. Here the average factor is also about 0.5. However, the factor varies across vowels and formants, and the variation is considerably greater for models that include teeth (0.2–0.9) as compared to models without teeth (0.3–0.7).

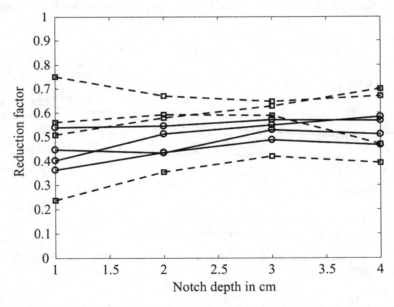

Fig. 7. Reduction factors for the /a/-resonators vs. notch depth. Solid lines: without teeth; dashed lines: with teeth. The four solid lines and the four dashed lines correspond to the four formants.

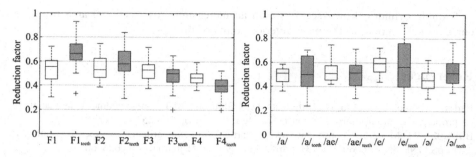

Fig. 8. Calculated length reduction factors for all measured conditions sorted by formants (left) and by vowels (right). Data for the resonators with teeth are shown as the gray boxplots.

4 Discussion and Conclusion

Lindblom et al. (2007) showed that it is possible to reproduce the formant frequencies of a uniform cylindrical tube with a notch at the lip opening with a corresponding unnotched tube, whose length is reduced by about 50% of the notch depth. Here we examined to what extent this finding holds for other resonator shapes beyond a uniform cylindrical tube and for the case that a row of teeth is included in the models. Across all conditions (vowel, formant index, notch depth), the average reduction factor was about 0.5 (50%), in accordance with Lindblom et al. (2007). However, the variation of this factor is considerably greater for resonators with teeth compared to resonators without teeth. Hence, for more realistic vocal tract shapes, the error can become relatively high when the notched lip region is replaced by a cylindrical tube section with half the length of the notch. The greatest variation of reduction factors was obtained for the vowel /e/ with teeth. So it seems that the impact of the teeth on the variation of the reduction factor is greatest for where the cross-sectional areas of the anterior vocal tract are small.

An aspect that was not analyzed here is the effect of a notch on the bandwidths of the formants. A notched lip opening has a bigger radiating surface and hence a bigger radiation impedance than an unnotched lip opening for the same vowel. Therefore, notching can be expected to increase the bandwidths of the formants more as compared to an unnotched case for which the formant frequencies are the same. So, using a shortened unnotched tube to replace a notched lip opening might underestimate formant bandwidths. However, that needs to be explored in detail in future work.

In conclusion, our data suggest that substituting the notch region of a notched tube model by a cylindrical tube section with a length of half the notch depth provides formant frequencies that are roughly equal to those of the notched resonator. The alternative approximation of the lip horn by a fixed-length tube section of 11 mm, as suggested by Badin et al. (1994), will not be optimal, because our data show that the length of the substitute tube section should vary with the notch depth. In future projects, it might be interesting to explore whether the 3D lip region of a resonator can be more accurately represented as a horn-shaped tube section in 1D tube models (instead of a single cylindrical tube section) with an increasing cross-sectional area from the corners of the mouth to the most anterior points of the lips.

References

Badin, P., Motoki, K., Miki, N., Ritterhaus, D., Lallouache, M.T.: Some geometric and acoustic properties of the lip horn. J. Acoust. Soc. Jpn. (E) 15(4), 243–253 (1994)

Birkholz, P.: Modeling consonant-vowel coarticulation for articulatory speech synthesis. PLoS ONE 8(4), e60603 (2013). https://doi.org/10.1371/journal.pone.0060603

Birkholz, P., Jackèl, D.: Influence of temporal discretization schemes on formant frequencies and bandwidths in time domain simulations of the vocal tract system. In: Proceedings of the Interspeech 2004-ICSLP, Jeju, Korea, pp. 1125–1128 (2004)

Engwall, O.: Combining MRI, EMA and EPG measurements in a three-dimensional tongue model. Speech Commun. 41, 303–329 (2003)

Flanagan, J.L.: Speech Analysis, Synthesis and Perception. Springer, Heidelberg (1965)

Fleischer, M., Mainka, A., Kürbis, S., Birkholz, P.: How to precisely measure the transfer function of physical vocal tract models by external excitation. PLoS ONE 13(3), e0193708 (2018). https://doi.org/10.1371/journal.pone.0193708

Kitamura, T., Takemoto, H., Adachi, S., Honda, K.: Transfer functions of solid vocal-tract models constructed from ATR MRI database of Japanese vowel production. Acoust. Sci. Technol. 30(4), 288–296 (2009)

Lindblom, B., Sundberg, J., Branderud, P., Djamshidpey, H.: On the acoustics of spread lips. Proc. Fonetik TMH-QPSR 50(1), 13–16 (2007)

Lindblom, B., Sundberg, J., Branderud, P., Djamshidpey, H., Granqvist, S.: The Gunnar Fant legacy in the study of vowel acoustics. In: Proceedings of the 10ème Congrès Français d'Acoustique, Lyon, France (2010)

Lindblom, B., Sundberg, J., Branderud, P., Djamshidpey, H.: Articulatory modeling and front cavity acoustics. Proc. Fonetik TMH-QPSR 51(1), 17–20 (2011)

Slaj, M., Spalj, S., Pavlin, D., Illes, D., Slaj, M.: Dental archforms in dentoalveolar Class I, II and III. Angle Orthod. 80(5), 919–924 (2010)

Stavness, I., Gick, B., Derrick, D., Fels, S.: Biomechanical modeling of English /r/ variants. J. Acoust. Soc. Am. 131(5), EL355–EL360 (2012)

Speech Acquisition

Production of Neutral Tone on Disyllabic Words by Two-Year-Old Mandarin-Speaking Children

Jun Gao and Aijun Li[✉]

Institute of Linguistics, Chinese Academy of Social Sciences, Beijing, China
9894575@qq.com, liaj@cass.org.cn

Abstract. This study examined the production of neutral tone in disyllabic words by two-year-old Mandarin-speaking children. The results showed that children were fully aware of the neutral tone sandhi rule phonologically at the age of two. However, they cannot phonetically produce neutral tone well. In particular, children made off-standard production with higher pitch register, wider pitch range and longer duration, while made correct production with correct pitch pattern but the duration ratio between the initial syllable and the final syllable is slightly larger than the adults'. The difficulty of the neutral tone production is closely related to the type of the preceding tone and the coordination of articulation for disyllabic neutral tone words.

Keywords: Neutral tone · Disyllabic words · Production · Two-years-olds Mandarin

1 Introduction

In Standard Chinese, apart from the four distinctive lexical tones, there exist weak elements in terms of neutral tone (Chao 1922, 1979), which is a special type of tone sandhi in Mandarin phonology. The neutral tone sandhi rule indicates that the neutral-tone syllable loses its lexical tone and becomes weak and short. The pitch contour of the neutral-tone syllable depends on the preceding tone within the same prosodic word. Even though words in neutral tone only take up a very small portion in the adults' lexicon, they appear in a relative higher percentage in children's lexicon. Studies on phonological development of children usually focused on the acquisition of consonants, vowels and lexical tones, with little attention paid to the acquisition of neutral tone. Li and Thompson (1977) reported that children (Taipei, in Taiwan) usually substituted full tone for neutral tone, especially for the neutral-tone words with affix -zi/tsɣ/. Zhu (2002) examined the acquisition of neutral tone of Mandarin-speaking children in Beijing in a more detailed way. Since neutral tone is pitch-related, it is interesting to know how the acquisition of neutral tone is related to the acquisition of lexical tones. The literature on Mandarin lexical tone acquisition showed that lexical

© Springer Nature Switzerland AG 2018
Q. Fang et al. (Eds.): ISSP 2017, LNAI 10733, pp. 89–98, 2018.
https://doi.org/10.1007/978-3-030-00126-1_9

tone is acquired at around two years of age (Zhu 2002; Si 2006).[1] Hence, we were motivated to explore whether Mandarin-speaking children have acquired neutral tone at the age of two by examining the acoustic patterns of the disyllabic neutral-tone words in both F0 and durational dimensions. We also sought to discover how the acquisition of neutral tone is related to the acquisition of lexical tone and whether there are general principles in the acquisition of tonal combinations.

2 Neutral Tone in Mandarin

Neutral tone is related to both tone and stress system in Standard Chinese (Lu and Wang 2005). Neutral tone does not occur in the initial position of a word and is assumed to be associated with weak syllable, which is short and light. The disyllabic neutral-tone words follow a strong-weak stress pattern. The final syllable being perceived weak is mainly due to several factors: dropping of the original tone, reduced pitch range, and shorter duration (Li 2017). The duration of neutral-tone syllable is about 55%–66% of its preceding syllable (Li 2017). The perception of neutral tone is determined more by F0 than duration (Li and Fan 2015). Spectral tilt is also proved to play a role in perception, but it is less important than F0 and duration. Intensity of neutral tone syllable is not always weaker than that of full tone syllable (Lin and Yan 1980). Vowel reduction is not a reliable cue either, which is highly related to personal habit or dialect background (Lin 2012). The final part of the neutral tone syllable is likely to be voiceless and the voiceless initial part is likely to be voiced (Li 2017).

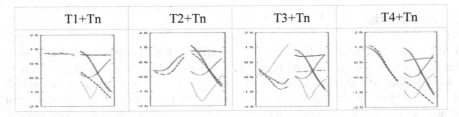

Fig. 1. The F0 contour patterns of all the two-tone combinations (dotted lines represent the neutral-tone combinations; solid lines represent full-tone combinations; Tn denotes four full tones and neutral tone). In the third panel, the high rising pitch contour on the first syllable is subjected to Tone3-Tone3 sandhi rule. (The F0 data are the averages of five males and five females normalized in Z-score. The X-axis is normalized duration.)

As shown by the dotted lines in Fig. 1, the neutral tone T0 is a falling tone behind Tone1 (T1, high-level tone), Tone2 (T2, high-rising tone), and Tone4 (T4, high-falling tone), while it is a mid-level tone behind Tone3 (T3, low-dipping tone). Regarding the

---

[1] Another study on lexical tone acquisition of our team showed that two-year-old Mandarin-speaking children still had around 13% tone errors and that the time of lexical tone acquisition was after 3.5 years of age.

falls on the neutral-tone syllable, it can be seen from Fig. 1 that there are some phonetic differences due to the offset of the preceding tone, i.e. a low fall following the high falling tone (the dotted line in the fourth panel), a mid fall following a high-rising tone (the dotted line in the second panel) and a half-low fall after a high-level tone (the dotted line in the first panel).

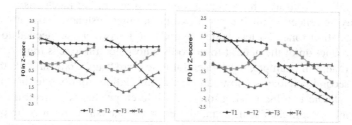

Fig. 2. The pitch contour patterns of all possible two-tone combinations on disyllabic prosodic words in Mandarin, full-tone combinations shown in the left panel, neutral-tone combinations shown in the right panel.

To have a clear picture of the pitch patterns of the tone combinations, we put all the full-tone combinations in one plot and all the neutral-tone combinations in the other plot as shown in Fig. 2. Again, it shows that the tonal contours of the second syllables are different in the two types of combinations: in full-tone combinations tonal contours of the second syllables maintain their citation form, but in neutral-tone combinations the second syllables exhibit a falling contour after T1, T2 and T4 and a mid level tone after T3. In addition, the pitch range of the second syllables is smaller in the case of neutral tone shown on the right panel than full tones on the left, and the pitch range of the first syllables in the neutral-tone words is larger than that in the full-tone words.

In Standard Chinese, neutral tone serves different lexical or morphosyntactic functions, as it can appear in about ten types of different contexts, such as some lexical items (lexeme type, such as /yi1 fu0/, "clothes"), stem-affix structures (affix type, such as -zi/tsʅ/), reduplications, locatives, directional complements, complement particle (de/tɤ/), aspect particles, particle (de/tɤ/), modal particles, and quantifier (ge/kɤ/). The last seven types are all related to grammatical morphemes. In the present study, we only focused on the acquisition of the first three types of neutral tone. These neutral-tone words are all disyllabic prosodic words.

3 Neutral-Tone Acquisition by Mandarin-Speaking Children

Zhu's study (2002) aimed to investigate the phonological development of normally developing Mandarin-speaking children. The study investigated 129 Mandarin-speaking children in Beijing aged from 1;6 to 4;6. The results showed that 57% of the youngest group (1;6–2;0, 21 children in total) could produce neutral tone and that except the error of the deletion of the affix -zi(/tsʅ/) almost all the errors were associated with pitch level and duration. Specifically, among the neutral-tone types like lexical item and reduplication, 93.4% of the errors were the substitution of the neutral tone by the citation tone. For example, the neutral-tone word "hair" /tou2 fa0/ was realized as /tou2 fa4/. According to Zhu, the high falling tone in /fa4/ – which was the underlying citation tone of the syllable /fa/ – substituted the neutral tone which should have been a mid fall with much shorter duration. But from our point of view, this substitution also could be the result of lifting the onset of the pitch contour and widening the pitch range, with a mid fall realized as a high falling tone. As a result, the mechanism underlying the error may be related to pitch register and pitch range instead of substitution by the underlying citation tone. To test which hypothesis is correct, we needed to examine children's production of the neutral-tone words whose neutral tone is transformed from a rising full tone. For example, for the neutral-tone word "sun" /tai4 ia.· ˙0/,, the neutral tone of the second syllable is a low fall after a high falling tone while the full tone of the second syllable is a high-rising tone2. If children produce the neutral tone of the second syllable as a high fall, it means the error mechanism is raising the pitch level and widening the pitch range instead of substitution of the neutral tone with a citation tone (high rising tone). More detailed analyses were needed to testify what mechanism underlies children's production errors of neutral tone.

In Zhu's study, there were two age groups including two-years-olds: 1;6–2;0 group (21 children) and 2;1–2;6 group (24 children). Only 13 "weakly stressed syllables" were examined. And there were no acoustic analyses reported on the error types (deletion, pitch level and duration). To see whether and how the two-year-olds acquire neutral tone, the acoustic patterns of more neutral words and more children were needed.

4 Data

We selected two-year-old children's production data of disyllabic neutral-tone words from a large-sample corpus (CASS_CHILD_Word, Gao et al. 2013). The corpus was the picture-naming production data of Mandarin-speaking children in the urban area of Beijing. The selected data were annotated in two rounds by both authors, using Praat (http://www.fon.hum.uva.nl/praat/).

379 tokens of 97 disyllabic neutral-tone words produced by 60 Mandarin-speaking children were analyzed. All of them were two years old, aged from 2;0(01) to 2;0(30), 26 boys and 34 girls. The disyllabic neutral-tone words covered three types: the lexeme type (132 tokens), the reduplication type (66 tokens) and the affix type (181 tokens). The reduplication type contained two subtypes, the reduplicative words already in the lexicon (40) like "dad" (/pa4pa0/), "mom" (/ma1ma0/), "baby" (/pao3pao0/), "star" (/xi.· ˙1 xi.· ˙0/), and the reduplicative words created by reduplicating monosyllabic words (26). In the affix type, most words were the words with the affix zi (163). In total, the data consisted of 77 T1 + T0 (T0 meant neutral tone) words, 126 T2 + T0 words, 55 T3 + T0 words and 121 T4 + T0 words.

All neutral-tone words are categorized into three groups based on the goodness of neutral-tone production: Correct, Off-standard and Non-neutral-tone. The error of the first tone was annotated as well. Table 1 illustrates the distribution of the neutral-tone production. If calculated by tokens, 59.5% neutral tones are correctly produced, 33.5%

Table 1. Distribution of neutral tone production

		Tokens (%)	Children (%)
Correct		59.5% (225/379)	93% (56/60)
Error	Off-standard neutral tone	33.5% (127/379)	73% (44/60)
	Non-neutral-tone	7% (27/379)	35% (21/60)

were produced as Off-standard and 7% were produced as Non-neutral-tone. If calculated by population[2], the statistics were 93%, 73% and 35% respectively.

F0 and duration of each tone were extracted and manually checked by the authors. F0 data were then normalized into 10 points for later analyses.

5 Results

5.1 Pitch

40.6% (154/379) of the data were judged wrong, i.e. not produced as neutral tone. These errors were made by 47 children, 22 boys and 25 girls. The errors could be divided into two groups, off-standard neutral tone and non-neutral-tone. For the off-standard-neutral-tone errors, the pitch movement of the final syllable followed the neutral-tone sandhi rule, but the pitch register was higher and the pitch range wider, which made the final syllable perceptually strong. This meant that children were aware of the sandhi rule. For the non-neutral-tone errors, the pitch movement of the final tone did not follow the neutral-tone sandhi rule. In these errors, some final syllables were produced in the citation tone, which made final syllables being perceived metrically strong. Other final syllables were produced as a level tone being perceived relatively weak.

Off-Standard-Neutral-Tone Error

127 word tokens (produced by 44 children, 21 boys and 23 girls) were judged as soff-standard-neutral-tone errors (82.5% of 154 errors). Among them, there were 27[3] Tone1-initial words (28.1% of 96 Tone1-initial words), 57 Tone2-initial words (45.8% of 117 Tone2-initial words), 22 Tone3-initial words (42.3% of 52 Tone3-initial words) and 21 Tone4-initial words (18.4% of the 114 Tone4-initial words).

It could be seen that off-standard-neutral-tone errors were more likely to occur in Tone2-initial and Tone3-initial disyllabic neutral-tone words (the right panel in Fig. 3). Compared to the correct productions (the left panel in Fig. 3), in off-standard-neutral-tone errors, for Tone2-initial disyllabic words (T2 + T0), the pitch contour on the final

---

[2] For some children, they could only have correct production on some neutral-tone words. As a result, the total of the children of the three production types exceeded 60.

[3] This was counted based on children's real production of the first tone rather than on orthography. Children made substitution errors on the first tone, so the number of Tone 1 on the first syllable was counted according to both correct production and substitutions. All the numbers hereafter were counted based on children's real productions.

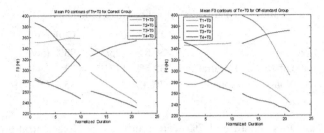

Fig. 3. The left panel is the mean F0 of the correct productions of Tn + T0 and the right panel is for those of the off-standard productions.

syllable had a higher onset and larger pitch range and for Tone3-initial disyllabic words (T3 + T0), the pitch contour on the final syllable had a higher register. Besides, T4 in T4 + T0 and T2 in T2 + T0 were not as steep as in the correct productions.

Non-Neutral-Tone Error

14 words (produced by 12 children, 6 boys and 6 girls) were produced as non-neutral-tones (9.1% of 154 errors) with the final syllable bearing level tone (the left panel in Fig. 4) despite whatever the initial tone was. The erroneously produced words were shirt, insect, wheel, baby, older sister, ear, moon, place and leopard. Of the 14 words, 6 words were produced with the first-tone substituted as Tone1.

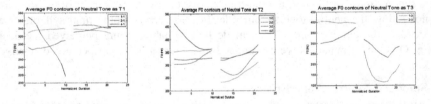

Fig. 4. The pitch contour patterns of the non-neutral-tone error with the final syllable bearing level tone Tone1 (left), high rising tone Tone2 (mid) and low rising tone Tone3 (right)

10 words (produced by 8 children, 5 boys and 3 girls) were produced as non-neutral tones (6.5% of 154 errors) with the substitution of the neutral tone by the citation Tone2 (the mid panel in Fig. 4), despite whatever the initial tone was. The reason why children produced neutral tone as Tone 2 might be that the final syllables of most of the 10 words (i.e. *cherry, lantern, clothes, steamed bun, sun*) carry a high rising tone when produced in isolation.

3 words (produced by 3 children, 1 boy and 2 girls) were produced (1.9% of 154 errors) with the neutral tone produced as Tone 3 (the right panel in Fig. 4). The reason for this wrong production might be that children added pragmatically a final successive boundary tone to the neutral tone. A fall neutral tone followed by a low rise made it perceived as Tone3. The initial tones of the three words included two in Tone1 and one in Tone2.

Errors on the Initial Tone

Fewer errors occurred on the initial tones. There were two kinds of errors in the initial tones. One was the off-standard production and the other was substitution. The initial Tone2 and Tone4 were more likely to be subjected to off-standard production and substitution as Tone1.

Even though children produced the initial tone wrongly into another tone, on most occasions they had the right neutral tone (or at least the right pitch contour) on the final syllable according to the 'new' initial tone. As such, in the off-standard cases of the initial tone, children could still produce the neutral tone on the final syllable correctly. These indicated that children were full aware of the neutral tone sandhi rule.

In these disyllabic words, the initial Tone2 was rarely produced wrongly as Tone3 and the initial Tone3 was rarely produced wrongly as Tone2. These were different from the cases of the monosyllabic tones where Tone2 and Tone3 are the frequent substitution targets for each other (according to the results of another study of ours).

Correct Productions of Neutral Tone

Figure 5 showed the pitch patterns of children's correct production of neutral tone grouped by the first tone. The mean contours were shown in the left panel of Fig. 3. Table 2 presented the maxima and minima of mean F0 of the first syllables and the neutral-tone syllables for correct and off-standard neutral-tone words (see Fig. 3).

In terms of neutral-tone syllables, the tonal range of the off-standard neutral-tone tokens was 48 Hz wider and the tonal register was 9 Hz higher than those of the correct neutral-tone tokens. In terms of the first syllable, the tonal range of the correct neutral-tone tokens was 54 Hz wider and the tonal register was 9 Hz higher than those of the off-standard neutral-tone tokens. The pattern of the Correct group was similar to the adults' production patterns (see Fig. 2).

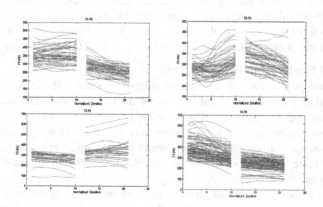

Fig. 5. Correct production of neutral tone grouped with the first tone.

Table 2. Mean F0(Hz) of Correct and Off standard productions.

	CorrSyll1	CorrNeu	OffSyll1	OffNeu
F0max	387	354	351	399
F0min	247	230	265	227
F0average	317	292	308	313
F0range	140	124	86	172

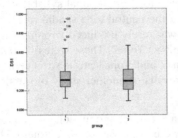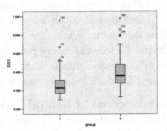

Fig. 6. Duration distribution for first syllable (DS1) and neutral tone syllable (DS2) for two groups. Group1 stands for the correct productions and group2 for the off-standard neutral-tone productions.

5.2   Duration

Duration distributions of the Correct and Off-standard groups were plotted in Fig. 6.

Correct Production of Neutral Tone

For correct productions of neutral tone, the mean duration of the first syllable was 331 ms (Sd. 129 ms) and that of the second syllable was 249 ms (Sd. 106 ms). The second syllable was shorter than the first syllable. On average, the duration of the final syllable was 0.86 of the first syllable. The ratio was larger than the ratio in adults (0.55–0.66) between the initial and the final syllables in disyllabic neutral-tone words.

Off-Standard Production of Neutral Tone

For off-standard productions, the mean duration of the first syllable was 324 ms (Sd. 130 ms) and that of the second syllable was 394 ms (Sd. 165 ms). The second syllable was longer than the first syllable. The duration of the first syllable in incorrect productions was similar to that in correct productions. But the second syllable in incorrect productions was much longer than that in correct productions. On average, the duration of the final syllable was 1.43 of the initial syllable. It meant that the final syllable was much longer than the first syllable.

ANOVA analysis showed that the durations of the first syllable were not significantly different between two groups ($F = 0.214$, $p = 0.64$), but the durations of the neutral tone syllable were significantly different between two groups ($F = 0.050$, $p = 0.00$).

5.3   Summary

Out of 97 words in the experiment, "ear" /ɻr3tou0/ and "eye" /ya.ˑˈ3tɕi.ˑˈ0/ are more likely to be produced with off-standard-neutral-tone error, as the pitch on the final syllable had a higher register. At the same time, "pomegranate"/ɭ 2liɭu0/, "grape"/p‿u2t‿ao0/, "hair"/t‿ou2fa0/ and "skirt"/tɕ‿i .ˑˈ2ts 0/ are more likely to be produced with off-standard-neutral-tone error, as the pitch on the final syllable had a much higher onset and wider pitch range.

In another study (Gao et al. 2017) on the acquisition of lexical tones by Mandarin-speaking children, we showed that the difficulty ranking of lexical tones on mono-syllabic words is T3 > T2 T1, > T4. Low-dipping tone is the most difficult tone, followed by rising and level tone; falling tone is the easiest tone. T3 is commonly mispronounced as T2 or produced in a non-canonical way. T2 is usually produced wrongly as T3. T1 is usually produced wrongly as T4 or T2. T4 is usually produced wrongly as T1 or produced in a non-standard way.

Comparing the production patterns of neutral tone and lexical tones, we concluded that children were fully aware of the pitch movement patterns in neutral tone at two years of age. They knew how the pitch of the neutral tone was going based on the preceding tone, because even when they incorrectly produced the previous tone into another tone, they produced the right pitch contour on the neutral tone in accordance with the 'new' preceding tone. And children also could apply the neutral-tone sandhi rule to the new reduplicated words which they had never heard, which were created by reduplicating a monosyllabic word.

Even though they are fully aware of the neutral-tone rule phonologically, they cannot produce it well enough phonetically. Children still need improvement on pitch register, pitch range and syllable duration in their productions of neutral tone. One reason might be in articulation. Children's articulatory organs and the coordination between these organs have not reached the level of maturity as seen in adults. They tend to produce higher pitch onset and pitch register, wider pitch range and lengthened duration than adults in neutral-tone syllables. Another reason might be due to the fact that the child-directed speech is slower with higher register and wider pitch range.

The reason that Tone2(LH)-initial and Tone3(LLH)-initial neutral-tone combinations are more prone to have a higher onset or register and wider pitch range of the pitch on the final syllable might be that the H target of the initial tone is realized on the second syllable. The peak delay is quite common when neutral tone is involved (Li 2003; Chen and Xu 2006).

In our data, compared to errors in consonants and vowels, lexical tones on the first syllable had fewer errors. This means that lexical tones are acquired earlier than consonants and vowels. For lexical tones, even though children still made errors on monosyllabic syllable (according to the results of another study of ours) and some on the initial tone, they generally showed good awareness of the neutral-tone rule. It seems that rules are acquired earlier than tones. This may be due to the fact that the domain within which neutral-tone sandhi rule applies is prosodic word, to which children are very sensitive in their language acquisition process.

Acknowledgements. The study is supported by two grants from the Key Project of National Fund of Social Sciences, No. 15ZDB103 and No. 13CYY025.

References

Chao, Y.R.: Gwoyeu Romatzyh or the national romanization. In: Wu, Z.J. (ed.) Linguistic Essays by Yuenren Chao, pp. 61–72. The Commercial Press, Beijing (1922)

Chao, Y.R.: Beijing Kouyu Yufa (A Grammar of Spoken Chinese). The Commercial Press, Beijing (1979)

Zhu, H.: Phonological Development in Specific Contexts: Studies of Chinese-Speaking Children (Child Language and Child Development). Multilingual Matters Ltd., Clevedon (2002)

Si, Y.: A case study of the acquisition of speech sounds by a Putonghua-speaking child. Contemp. Linguist. **8**(1), 1–16 (2006)

Lu, J.L., Wang, J.L.: Guanyu qingsheng de dingjie (Analysis on the nature of neutral tone. Contemp. Linguist. **7**(2), 107–112 (2005)

Li, A.: Phonetic correlates of neutral tone in different information structures. Contemp. Linguist. **19**(3), 348–378 (2017)

Li, A., Fan, S.: Correlates of Chinese neutral tone perception in different contexts. In: Proceedings of the 18th International Congress of Phonetic Sciences (ICPhS 2015), Glasgow, Scotland (2015)

Li, C.N., Thompson, S.A.: The acquisition of tone in Mandarin-speaking children. J. Child Lang. **4**(2), 185–199 (1977)

Lin, M.C., Yan, J.Z.: Beijinghua Qingsheng de Shengxue Xingzhi. Dialect **3**, 166–178 (1980)

Lin, M.: Experimental Study on Chinese Intonation. China Social Sciences Press, Beijing (2012)

Li, Z.: A perceptual account of asymmetries in tonal alignment. In: Kadowaki, M., Kawahara, S. (eds.) Proceedings of the North East Linguistic Society 33, pp. 147–166. UMass GLSA, Amherst (2003)

Chen, Y.Y., Xu, Y.: Production of weak elements in speech evidence from F0 patterns of neutral tone in standard Chinese. Phonetica **63**, 47–75 (2006)

Gao, J., Li, A., Xiong, Z., Shen, J., Pan, Y.: Normative database of word production of Putonghua-speaking children - Beijing articulation norms project: CASS_CHILD_WORD. In: Proceedings of Oriental-COCOSDA 2013 (2013)

Gao, J., Li, A., Zhang, Y.: The production of tones in monosyllabic and disyllabic words by Mandarin-speaking children. In: Proceedings of the ISSP, OC, Tianjin, pp. 16–19 (2017)

English Lexical Stress Production by Native Speakers of Tibetan and Uyghur

Dan Hu[1], Hui Feng[1,2,3(✉)], Yingjie Zhao[1], and Jie Lian[1]

[1] School of Foreign Languages and Literature, Tianjin University, Tianjin, China
fenghui@tju.edu.cn
[2] Research Center for Linguistic Sciences, Tianjin University, Tianjin, China
[3] Tianjin Key Laboratory of Cognitive Computing and Application, Tianjin, China

Abstract. This study studies the production of English lexical stress by native speakers of Tibetan and Uyghur, and the factors that may affect stress assignment. Thirty subjects in their twenties participated, with 10 native speakers (gender balanced) for each language, i.e. native speakers of Uyghur (NSUs), native speakers of Tibetan (NSTs) and native speakers of American English (NSAs). A total of 4,000 tokens are collected, judged and analyzed. Results indicate that: (1) Consistent with the prediction of Stress Typology Model, less negative transfer has been observed in NSTs than in NSUs in stress production. (2) Compared with NSAs, NSUs and NSTs employ different acoustic features when assigning stress. (3) Stress positions affect the accuracy of stress production by NSUs, and also the acoustic features of NSTs and NSUs when assigning stress. A speech-final lengthening effect is observed. (4) Syllable structures have little effect on the accuracy of stress production.

Keywords: English stress production · Stress Typology Model
Tibetan · Uyghur · Stress position · Syllable Structure

1 Introduction

Previous studies [1–4] have reported that four acoustic features (AFs), namely, fundamental frequency (F0), intensity (INS), syllable duration (DUR), and vowel quality (schwa), are employed to realize English stress. In the past 20 years, increasing attention has been paid to lexical stress acquisition. Altmann [5] has investigated English lexical acquisition by seven distinct first language (L1) groups (Arabic, Chinese, French, Japanese, Korean, Spanish, Turkish) and results verified Stress Typology Model (STM). STM predicts that speakers of non-stress L1 would have the best performance for English lexical stress, speakers of L1 s with predictable stress should meet with the greatest difficulties, and the performance of L1 groups with non-predictable stress would be among them [5]. Except for stress typology of L1, such factors as word class, stress position and syllable structure [6, 7] are reported to have effect on English stress acquisition.

© Springer Nature Switzerland AG 2018
Q. Fang et al. (Eds.): ISSP 2017, LNAI 10733, pp. 99–108, 2018.
https://doi.org/10.1007/978-3-030-00126-1_10

To test whether speakers of stress language and those of tone language have different performance in lexical stress production, this study carries out a study on the production of English lexical stress by native speakers of Tibetan and Uyghur. An increasing number of studies has reported that Uyghur is a stress-accent language with free stress [8–10], which suggests that Uyghur has the same stress typology as English. Different from English, duration is a robust cue to stress in Uyghur, and F0 is not used to distinguish stressed syllables from unstressed ones within one word [8–10].

According to Qu [11] and Jin [12], Tibetan includes three main dialects: Ü-Tsang, Khams, and Amdo. In this study, Tibetan is referred to as Ü-Tsang, which has four tones, namely, high level, high falling, low rising and low dipping [13]. Tibetan tones are subject to syllable structure constraints, that is, a syllable closed by a nasal consonant ((C)VN) can bear all four tones, open syllables with monophthongs ((C)V) and diphthongs ((C)VV) can bear only high level tone and low rising tone, and a syllable closed by a voiceless stop ((C)VG) can only bear high falling tone and low dipping tone [14].

Uyghur, a stress-accent language with free stress, may impede the second language (L2) lexical stress acquisition by its native speakers, while Tibetan, a tone language, may bring little impediment to its speakers in the acquisition of L2 lexical stress. This study aims to evaluate the English lexical acquisition by native speakers of Tibetan and Uyghur, as well as to find out possible factors that may affect their stress assignment.

Three research questions are to be answered in this study: (1) Whether or not the results of the stress production by native speakers of Tibetan and Uyghur are consistent with the predictions of STM, that is, Tibetan speakers would perform stress better than Uyghur. (2) Compared with American, what acoustic features native speakers of Tibetan and Uyghur take in stress realization? Do the speakers of Tibetan and Uyghur tend to decrease these features of the unstressed syllable, or raise them of the stressed syllable, or widen the range of such features between the stressed and the unstressed? (3) Whether stress positions and syllable structures have any effect on the English stress production by native speakers of Tibetan and Uyghur?

2 Methods

2.1 Subjects

Thirty subjects participated, with 10 subjects (gender balanced) for each language, i.e. 10 native speakers of Uyghur (NSUs), 10 native speakers of Tibetan (NSTs), and 10 native speakers of American English (NSAs). They are divided into two groups: the Uyghur-American group (10 NSUs and 10 NSAs) and the Tibetan-American group (10 NSTs and the same 10 NSAs). NSUs and NSTs are English learners in their twenties who have learned English for at least 10 years, and NSAs are native speakers of American English. All subjects were right-handed, reported normal hearing and speaking.

2.2   Stimuli

Pseudo-words are used to minimize the familiarity effect that will inflate the scores [7]. Stimuli for the Uyghur-American group are 72 disyllabic pseudo-words whose syllabic structures conform to the phonotactics of both Uyghur and American English. In Uyghur, syllabic structures include V, VC, CV, CVC, VCC, CVCC, and no consonant clusters are allowed in disyllabic words [15]. Besides, constraints on the syllabification rule in disyllabic words [16] prescribe that when there is one consonant between vowels, this consonant acts as the onset of the second syllable, when two consonants, the first acts as the coda of the first syllable, and the second as the onset of the second syllable, when three consonants, the first two act as coda in the first syllable and the third is the onset of the second syllable, and when there is no consonant in-between, the two vowels act as the rhymes of two syllables. Therefore, nine disyllabic structures are shared in Uyghur and English: V.CV, V.CVC, VC.CV, VC.CVC, CV.VC, CV.CV, CV.CVC, CVC.CV, CVC.CVC.

The shared vowels by English and Uyghur were used in the stimuli, and the consonant used in the words is *d*. Each word is put in a carrier sentence "I say _____ again". For example, *a.di* was one of the four pseudo-words for the V.CV structure. Each of these 36 words was to be read twice, with one time stressed on the first syllable and the other on the second. Subjects were informed that they were reading pseudo English words, and were asked to read all those 72 words (36 * 2) in the carrier sentence, and totally 1440 (9 types * 4 words * 2 times * 20 subjects) tokens were collected from the production experiment in the Uyghur-American group.

Similarly, stimuli for the Tibetan-American group are 64 disyllabic pseudo-words whose syllabic structures conform to the phonotactics of both Tibetan and American English, for example "*i·dam*". Table 1 displays the 16 shared structures used to coin disyllabic pseudo-words, combined by "CV", "CVV", "CVN" and "CVG". With four words for each structure, 2560 (16 types * 4 words * 2 times * 20 subjects) tokens were collected from the production experiment in the Tibetan-American group.

Table 1. Disyllabic structures shared between Tibetan and English

	CV	CVV	CVN	CVG
CV	CV.CV	CV.CVV	CV.CVN	CV.CVG
CVV	CVV.CV	CVV.CVV	CVV.CVN	CVV.CVG
CVN	CVN.CV	CVN.CVV	CVN.CVN	CVN.CVG
CVG	CVG.CV	CVG.CVV	CVG.CVN	CVG.CVG

Notes: *N* stands for a nasal consonant and *G* a voiceless stop.

2.3   Procedure

Reading materials, i.e., the pseudo-words for each group, for the production task were randomly shown on the computer screen, with the intended stressed syllable in blue and the unstressed one in grey. Each subject finished the corresponding task independently. The speech was recorded directly onto a hard drive disk at a 16 kHz 16 bit sampling rate, and saved as wav files for further evaluation and analysis.

Another 30 well-trained native English speakers evaluate the recordings to judge whether the stressed position is at the initial or final syllable. The perception results were then compared with the assigned task to check whether the subjects had produced the stressed syllable as required. Disparities for the same recording among the 30 raters were decided by the acoustic features analysis. For example, if the first syllable is asked to be stressed, only when all AFs of the first syllable is larger than the second one will we decide the subject has stressed the word correctly. At last, each subject gets an average accuracy rate for their performance of stress production. For further analysis, three acoustic features (fundamental frequency (F0), intensity (INT) and duration (DUR)) of each token are extracted and normalized.

2.4   Normalization

To eliminate individual physiological differences, especially those between genders, acoustic features are normalized based on a z-score transform since the data meet Norm-distribution pattern. This study specified the highest and lowest values of an individual's overall speaking range and defined the phonetics of acoustic features (F0, INS, and DUR) relative to these points, which is shown in Formula (1).

$$M = [(m - Min)/(Max - Min)] * 100\% \tag{1}$$

where "M" is the normalized mean for each syllable, "m" is the original mean for each syllable, "Max" is the standard deviation above the mean of one subject's overall acoustic features, and "Min" is two standard deviations below the mean of the same subject's overall acoustic features. For example, if we get M = 69.38 as the normalized F0 mean of Subject 1's stressed syllable, we can say Subject 1 produces this syllable at the place of 69.38% in his personal pitch range.

3   Results, Findings and Discussion

Accuracy rate and acoustic features are analyzed to evaluate performance. The acoustic data were obtained from the correctly produced tokens. Results of the production experiments are presented from three aspects: overall results analysis, effects of stress positions and syllable structures.

3.1   Overall Results Analysis

In the Uyghur-American group, NSAs have an obvious higher average accuracy of stress production than NSUs (97.67% VS 80.00%, P = .003). In the Tibetan-American group, NSAs outperform NSTs with higher average accuracy in stress production (97.50% VS 84.77%, P = .002). In both groups, NSAs can locate stress with a much higher accuracy rate. For each group, with the mean normalized acoustic features of the stressed syllable as the top line and those of unstressed syllable as the bottom line, mean AFs of each group are shown in Fig. 1.

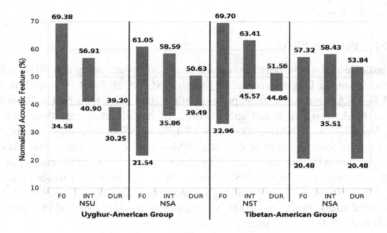

Fig. 1. Normalized mean values of AFs

As shown in the left panel of Fig. 1, i.e. the Uyghur-American Group, Uyghur speakers have different performance in AFs with American. NSUs produce the stressed syllable with higher F0 (69.38% VS 61.05%, P = .042) and shorter duration (39.20% VS 50.63%, P = .000) than NSAs, but no statistical difference with NSAs in intensity (P = .810). Second, NSUs produce the unstressed syllable with higher F0 (34.58% VS 21.54%, P = .000) and shorter duration (30.25% VS 39.49%, P = .001) than NSAs, but no statistical difference with NSAs in intensity (P = .135). Third, a narrower range is found with NSUs than with NSAs in F0 (34.80% VS 39.51%, P = .023), but no statistical difference with NSAs in intensity (P = .114) and duration (P = .574).

In the Tibetan-American group (the right panel of Fig. 1), Tibetan speakers also differ from American in AFs. Firstly, NSTs produce the stressed syllable with higher F0 (69.70% VS 57.32%, P = .000) and stronger intensity (63.41% VS 58.43%, P = .005) than NSAs, but no statistical difference with NSAs in duration (P = .225). Secondly, NSTs produce the unstressed syllable with higher F0 (32.96% VS 20.48%, P = .000), stronger intensity (45.57% VS 35.51%, P = .000) and longer duration (44.86% VS 20.48%, P = .001) than NSAs. Thirdly, NSTs and NSAs employ almost the same F0 ranges (36.74% & 36.84%, P = .800), but NSAs use wider range in intensity (22.92% VS 17.84%, P = .015) and duration (33.36% VS 6.70%, P = .001).

Compared with NSTs, NSUs has a wider range in duration (8.95% VS 6.70%) and a narrower range in F0 (34.80% VS 36.74%). This might be influenced by L1 phonology: Uyghur is a stress-language in which duration is a robust cue to stress [8–10], Tibetan is a tone-language in which pitch (F0) pattern distinguishes tone types. In summary, both NSUs and NSTs perform worse than NSA. Besides, compared with NSAs, NSUs and NSTs employ AFs with different strategies to realize stress. NSUs tend to raise F0 and shorten duration for both syllables in a disyllabic word, and have a narrower range of F0 between the stressed one and the unstressed one. NSTs tend to raise F0 and intensity for both syllable, elongate the duration of unstressed syllable, and have narrower range of intensity and duration between the stressed syllable and the unstressed one than NSAs.

3.2 Stress Position Effect

Results indicate that stress position has influence on the accuracy and AFs' values of stress production. In Uyghur-American group, ANOVA analysis (P = .025 in NSUs; P = .063 in NSAs) indicates stress position effect only exist with NSUs, though both NSUs and NSAs produce initial stress with a higher accuracy rate than final stress (92.22% VS 67.78% in NSUs, 100% VS 95.33% in NSAs). Besides, results of normalized acoustic features indicate that both NSUs and NSAs realize initial stress with wider ranges of F0 and intensity than final stress, with statistical significance (P < 0.05). As shown in Fig. 2, NSUs realize initial stress with a wider range of F0 (43.82% VS 24.08%, p = .000) and higher intensity (20.54% VS 11.39%, p = .040) than with final stress. NSAs realize initial stress with a wider range of F0 (48.83% VS 30.33%, p = .000) than final stress.

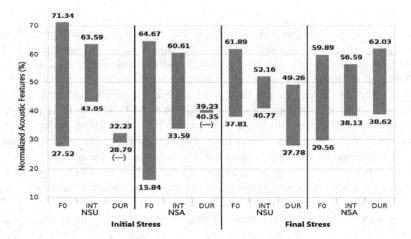

Note: "-" means AF value of the stressed syllable are smaller than the unstressed one.

Fig. 2. Normalized AFs in different stress positions for the Uyghur-American Group.

As for duration, an obvious speech-final lengthening effect is observed in Fig. 2. In the realization of initial stress (left panel), the DUR columns are in the negative direction (with a minus sign underlie the columns), which means the unstressed syllable has a longer duration than the stressed syllable. In the section of final stress (right panel), the columns of DUR are in the positive direction, which means the stressed syllable has a longer duration than the unstressed syllable. Such phenomenon indicates that both NSUs and NSAs produce the second syllable with longer duration regardless of the stress position. Such speech-final lengthening effect has been reported in previous studies [7, 17, 18].

In Tibetan-American group, both NST and NSA produce initial stress with higher accurate rate (89.06% VS 80.47% in NSTs, 98.59% VS 96.41% in NSAs), but neither gets support from ANOVA test (P = .291 in NSTs; P = .215 in NSAs).

As for the effect of stress positions on acoustic features, both NSTs and NSAs realize initial stress with wider ranges of F0 and intensity, and a shorter range of duration than final stress with statistical significance (P < .05). As shown in Fig. 3, NSTs' initial stress has longer columns of F0 (48.22% VS 25.25%, p = .000) and INT (28.34% VS 7.35%, p = .000), but has a shorter column of DUR (16.08% VS 29.47%, p = .000). NSAs' initial stress also has longer columns of F0 (45.70% VS 29.01%, p = .000) and INT (34.82% VS 11.02%, p = .000) than the final stress, but also has a shorter column of DUR (4.28% VS 26.41%, p = .000). Although the second syllable is not always the longer syllable in the Tibetan-American group, all subjects produce final stress with a longer range between the stressed one and the unstressed one than with the initial stress, which also indicates a slight speech-final lengthening effect.

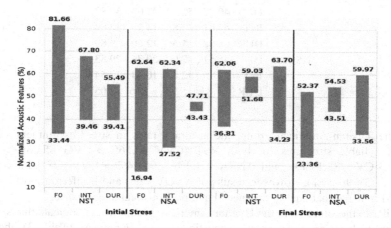

Fig. 3. Normalized AFs in different stress positions for the Tibetan-American Group.

To summarize, stress positions affect both accuracy and acoustic features for NSAs, NSTs and NSUs. Firstly, the stress position affect NSUs to assign initial stress with higher accuracy. At the same time, all subjects produce initial stress with a wider range of F0 and intensity. Lastly, a speech-final lengthening effect is observed, and such effect is more obvious in the Uyghur-American group.

3.3 Syllable Structure Effect

In the Uyghur-American group, although the accuracy rates of locating stress at the initial or final position on different syllable structures are different for NSUs (V 91.00% > VC 89.33% > CVC 80.00% > CV 72.86%) and NSAs (V 100.00% > VC 98.00% > CVC 97.33% > CV 97.14%), ANOVA test and Bonferroni post-hoc test indicate that such difference has no statistical significance (P > .05). Since syllable structures have little influence on accuracy, and NSUs locate stress less accurately than NSTs (80.00% VS 84.77%, see Sect. 3.1), thus, a conclusion from the experiment is that speakers of tone language (NSTs) would meet with less negative influence than speakers of free stress language (NSUs) in stress assignment, which is consistent with STM's prediction.

However, results indicate that syllable structures have an effect on acoustic features. For NSAs, as shown in Table 2, the AFs range in different stressed syllables vary. ANOVA analysis and Bonferroni post-hoc tests suggest that only F0 (p = .003) is affected by syllable structures. Such effect is not obvious in intensity and duration. For NSUs, the AFs ranges in stressed syllables are different. However, ANOVA analysis and Bonferroni post-hoc tests show that syllable structures have an effect on F0 (p = .015) and duration (p = .000), and no such effect has been observed on intensity.

Table 2. AFs range (%) of stressed syllables in Uyghur-American Group

Subjects	AFs	V	VC	CV	CVC
NSAs	F0	40.27	52.58	37.03	35.82
	INT	43.01	28.85	14.25	22.80
	DUR	24.08	1.59	12.60	9.89
NSUs	F0	45.51	42.00	30.76	29.83
	INT	34.59	12.83	10.64	17.54
	DUR	−1.83	15.98	−1.16	2.05

In the Tibetan-American group, the accuracy rates in stress assignment vary in different syllable structures for both NST (CVV 86.56% > CVG 85.63% > CVN 85.00% > CV 82.19%) and NSA (CVN 98.44% > CVV 97.81% > CVG 97.59% > CV 96.25%). However, results of ANOVA test and Bonferroni post-hoc test (P > .05) indicate that such differences have no statistical significance.

Similar to the situation in the Uyghur-American group, results indicate that syllable structures do have an effect on the acoustic features. For NSAs (Table 3), the AFs range in different stressed syllables vary. ANOVA analysis and Bonferroni post-hoc tests indicate F0 (p = .025) and duration (p = .000) are affected by syllable structures, but such an effect is not found on intensity. Compared with NSAs, NSTs differ in the AFs ranges in different stressed syllables. However, ANOVA analysis and Bonferroni post-hoc tests indicate that syllable structures have no effect on DUR, F0 or INT.

Syllable structures do have influence on acoustic features (especially on F0 and duration) of all subjects, with little effect on accuracy. ANOVA analysis and Bonferroni post-hoc tests show that syllable structures have effect on F0 and duration of NSUs and NSAs, and on duration of NSTs.

Table 3. AFs range (%) of stressed syllables in Tibetan-American Group

Subjects	AFs	CV	CVV	CVN	CVG
NSAs	F0	37.01	33.70	34.38	43.75
	INT	19.76	24.97	20.71	26.25
	DUR	8.89	18.86	25.47	8.17
NSTs	F0	39.12	35.67	38.52	33.73
	INT	14.74	18.22	16.55	21.86
	DUR	2.52	11.28	15.04	2.98

4   Conclusion

By conducting production experiments and analyzing results of accuracy rates and acoustic features (F0, intensity, and duration), this study has investigated whether language typology, stress positions and syllable structures affect English stress acquisition by native speakers of Tibetan and Uyghur. There are four major findings in this study. (1) Both NSUs and NSTs experience certain negative transfer in the acquisition of lexical stress, which is suggested by lower accuracy. NSTs meet less negative transfer than NSUs do, which is consistent with STM's prediction that tone language (in our experiment NSTs) would meet with less negative influence than speakers of free stress language (in our experiment NSUs) in stress assigning. (2) Compared with NSAs, NSUs and NSTs employ different acoustic features when assigning stress. NSUs tend to raise F0 and shorten duration for both syllables in a disyllabic word, and have a narrower range of F0 between the stressed and the unstressed. NSTs tend to raise F0 and intensity for both syllables, elongate the duration of unstressed syllables, and have a narrower range of intensity and duration between the stressed syllable and the unstressed ones than NSAs. (3) Stress positions affect the accuracy of NSUs and the acoustic features (especially F0 and intensity) of all subjects. A speech-final lengthening effect is observed in all subjects. (4) Syllable structures influence acoustic features, especially F0 and duration, though little influence is found with accuracy.

Such findings will provide empirical data for more effective teaching of English to minority groups. It will also be meaningful to analyze the acoustic features of those mis-assigned tokens to dig out more about possible factors (for example, F0, intensity, or duration) that may affect the acquisition of English tress. Besides, the role of vowel reduction in stress realization [19] can be analyzed to examine whether NSTs and NSUs tend to incorrectly reduce vowels in unstressed syllables.

Acknowledgements. The research is supported partially by Major Program of National Social Science Foundation of China (No. 15ZDB103), JSPS KAKENHI Grant (16K00297), National Natural Science Foundation of China (No. 61771333), Social Science Foundation of Tianjin, China (No. TJWW17-010), and Innovative Talent Training Project of Tianjin University (No. YCX17024).

References

1. Fry, D.B.: Experiments in the perception of stress. Lang. Speech **1**(2), 126–152 (1958)
2. Bolinger, D.L.: Pitch accent and sentence rhythm. In: Forms of English: Accent, Morpheme, Order, pp. 139–182. Hokuou, Tokyo (1965)
3. Lehiste, I.: Suprasegmental features of speech. In: Contemporary Issues in Experimental Phonetics, pp. 225–239. Elsevier Inc, Amsterdam (1976)
4. Beckman, M.E.: Stress and Non-stress Accent. Walter de Gruyter, Berlin (1986)
5. Altmann, H.: The Perception and Production of Second Language Stress: A cross-Linguistic Experimental Study. University of Delaware, Newark (2006)
6. Wayland, R., Landfair, D., Li, B., Guion, S.G.: Native Thai Speakers' Acquisition of English Word Stress Patterns. J. Psycholinguist. Res. **35**(3), 285–304 (2006)
7. Jangjamras, J.: Perception and Production of English Lexical Stress by Thai Speakers. University of Florida, Gainesville (2011)
8. Wang, H.: Perception and acoustic features of vowels' length in Uyghur. In: 9th Conference of Chinese Phonetics, Tianjin, pp. 7–13 (2013). (in Chinese)
9. Kadir, T.: Formal judgment on the Uyghur word stress. Lang. Transl. **4**, 38–46 (2015). (in Chinese)
10. Yakup, M., Sereno, J.A.: Acoustic correlates of lexical stress in Uyghur. J. Int. Phonetic Assoc. **46**(1), 61–77 (2016)
11. Qu, A.T.: Tibetan tones and tibetan tones' development. Stud. Lang. Linguist. **1**, 177–194 (1981). (in Chinese)
12. Jin, P.: Zangyu Jianzhi (Outline of Tibetan). Minzu Press, Beijing (1983). (in Chinese)
13. Zhou, J.W.: Textbook of Tibetan (Lhasa) Pinyin. Minzu Press, Beijing (2004). (in Chinese)
14. Hu, T.: A study of Tibetan (Lhasa) tones. Minor. Lang. Chin. **1**, 22–36 (1980). (in Chinese)
15. Zhang, Y.: Comparison of syllable structures between Mandarin and Uyghur. Xinjiang Univ. J. (Philos. Soc. Sci. Edn.) **1**, 105–108 (1998)
16. Azegour, A.: Introduction to Uyghur, p. 44. Mingzu University of China Press, Beijing (2006). (in Chinese)
17. Ueyama, M.: Phrase-final lengthening and stress-timed shortening effects in native speakers and Japanese learners of English. J. Acoust. Soc. Am. **98**(5), 2893 (1995)
18. Hu, D., Feng, H., Wu, T.: English stress acquisition by native speakers of Tibetan. In: 10th International Symposium on Chinese Spoken Language Processing, pp. 1–5. IEEE Press, New York (2016)
19. Saha, S.N., Mandal, S.K.D.: Phonetic realization of English lexical stress by native (L1) Bengali speakers compared to native (L1) English speakers. Comput. Speech Lang. **47**, 1–15 (2017)

Phonetics

C-V and V-C Co-articulation in Cantonese

Wai-Sum Lee[(✉)] [iD]

City University of Hong Kong, Tat Chee Avenue, Hong Kong, China
w.s.lee@cityu.edu.hk

Abstract. The present study investigates the co-articulation strength (CS) between the initial/final consonants and the neighboring vowels in Cantonese CV, VC, and CVC monosyllables, where C = [p t k] and V = [i a u]. EMA AG500 was used for recording the articulatory actions of the tongue and the lips during the test syllables. The findings based on the articulatory data collected from two male Cantonese speakers are as follows. First, CS is strong (*i*) between the initial [p-] and the following [i] or [u], but not [a], and (*ii*) between the final [-p] and a preceding vowel of any type. Second, CS is weak (*i*) between the initial [t-] and the following vowel of any type and (*ii*) between the final [-t] and the preceding [u], but not [i] or [a]. Third, CS is weak between the initial [k-] and the following [a], but not [i] or [u], however high between the final [-k] and the preceding [a]. In general, (*i*) the order of decreasing CS for both C-V and V-C co-articulation is when C = [p] > C = [k] > C = [t] and (*ii*) the degree of CS is higher in the VC than the CV context. The findings support the phonological structuring of the syllable, wherein the final consonant (or syllable coda), but not the initial consonant (or syllable onset), and the preceding vowel (or syllable nucleus) form the phonological unit of the rhyme.

Keywords: Cantonese · Co-articulation · Consonant · Vowel

1 Introduction

In real speech, concatenated speech sounds are uttered not as discrete segments but co-articulated units overlapping in time. The phenomenon of co-articulation has received attention from phoneticians and speech scientists. Earlier works have reported consonant context effects on vowels that (*i*) vowel formant frequencies are displaced away from the target frequencies as a result of consonant and vowel co-articulation [8, 10], and (*ii*) the extent of the consonantal influence on vowel formant frequencies is determined by factors, such as place of articulation, manner of articulation, and voicing characteristic of the consonants [10]. There have been a number of other studies of co-articulation of consonants and vowels in different languages through analysis of the patterns of vowel formant transition. Krull [5–7], by applying the Locus Equation (LE) (Formula 1), measure the co-articulation strength (CS, henceforth) between consonant and vowel in CV syllables based on the F_2 transitions between the vowel onset (F_2onset) and vowel center (F_2center).

© Springer Nature Switzerland AG 2018
Q. Fang et al. (Eds.): ISSP 2017, LNAI 10733, pp. 111–120, 2018.
https://doi.org/10.1007/978-3-030-00126-1_11

$$F_2onset = k * F_2center + c \qquad (1)$$

The equation expresses the linear regression of F_2onset on the y-axis against F_2center on the x-axis for determining C-V co-articulation when a given initial consonant is followed by various types of vowels. In the equation, k is the slope of the regression line and c the y-intercept of the slope. An increase in the k-value is taken to indicate an increase in the between-segment CS. Krull's LE data on C-V co-articulation between the Swedish initial voiced stops [b d g] and various types of the following vowels show that the order of decreasing CS is when C = [b] > C = [g] > C = [d]. Similar LE data have also been reported on the variations in F_2 pertaining to C-V co-articulation in relation to place of articulation of the initial consonants in other languages, such as English [13], Catalan [15], French [1], Thai, Arabic, and Urdu [12].

LE has also been applied to investigation of V-C co-articulation by measuring the F_2 transitions between the vowel center and vowel offset. A comparison of the LE data on the variations in F_2 for C-V and V-C sequences, where C = a voiced stop [b], [d], or [g] in English [11] and Persian [9], shows that in both languages the between-segment CS reduces for V-C co-articulation as compared to C-V co-articulation, when C = [d] and more even so when C = [b]. When C = [g], the between-segment CS for V-C co-articulation, relative to that for C-V co-articulation, increases in English, but decreases in Persian.

Co-articulation effects have also been investigated with respect to the compatibility of tongue position between consonant and vowel. Reported in [2] and [14], CS reduces when the neighboring consonant and vowel compete for the use of the tongue. The articulatory data on the Catalan consonants in [16] show that CS in relation to the antagonism of tongue position between consonant and vowel is weak for the alveolopalatal consonants but strong for the labial consonants, with the alveolar and velar consonants coming in between. In English [3, 4], the between-segment CS is weaker for the coronal consonants, including dental, alveolar, and postalveolar, due to a less varied tongue position, than the labial consonants.

To my knowledge, no published articulatory data on the co-articulation between consonant and vowel in Cantonese are currently available. The present study is an articulatory investigation of the co-articulation of the syllable-initial and syllable-final stop consonants [p t k] with different types of neighboring vowels in Cantonese monosyllables. It examines the variations in the tongue and lip positions at the C-V and V-C transitions to determine (i) the CS between the different types of stop consonants and vowels and (ii) the similarities and differences in CS between the syllable-initial/final stops and the neighboring vowels in the C-V and V-C contexts during the articulation of the Cantonese monosyllables.

2 Method

2.1 Test Materials

In Cantonese, the unaspirated stops [p t k] occur in both the initial and final positions of monosyllables. Table 1 presents a total of 25 test Cantonese monosyllables of the CV,

VC, or CVC structure, where the syllable-initial C- and syllable-final -C are one of the three stops [p t k] and V is one of the three corner vowels [i a u]. All the test syllables are meaningful monosyllabic words and are commonly used in daily communication in the Cantonese-speaking community in Hong Kong. The asterisk '*' in the table denotes the non-occurring syllables in Cantonese. A randomized reading list containing five repetitions of each test word was prepared for eliciting speech samples from Cantonese speakers.

Table 1. Test Cantonese monosyllables (*denoting non-occurrence in Cantonese).

CV			VC			CVC					
[pi]	[pa]	*	[ip]	[ap]	*	*	[pit]	*	[pat]	[pak]	[put]
[ti]	[ta]	*	[it]	[at]	[ut]	[tip]	[tit]	[tap]	[tat]	*	*
[ki]	[ka]	[ku]	*	[ak]	*	[kip]	[kit]	[kap]	*	[kak]	*

2.2 Speakers

Two male speakers, who were born and grew up in monolingual Cantonese-speaking families in Hong Kong, provided speech samples for this study. The two speakers were undergraduate students of the ages ranging from 18 to 22 with no history of speech and hearing difficulties.

2.3 Data Collection and Analysis

Electromagnetic Articulography (EMA) AG500 by Carstens of Germany was used to record and analyze the test materials. Synchronized digital recordings of the articulatory actions of the tongue and lips and the audio signals during the test syllables were performed. Throughout the recording, frequency receivers or sensors in small coils were fixed onto speaker's articulators, with (*i*) three sensors in equidistance on the tongue tip (TT), tongue middle (TM), and tongue back (TB) and (*ii*) two on the midpoints of the vermilion borders of the upper and lower lips. Three additional sensors on fixed locations include one on the nose bridge and the other two on the back of the ears, functioning as the reference points for tracking the head movements.

The articulatory data with respect to the positions of TT, TM, and TB and the positions of the upper and lower lips were extracted on the *x- and y*-coordinate planes, indicating the front-back and up-down positions of the tongue and the lips. By making reference to the waveforms and spectrograms of the synchronized acoustic signals, the tongue and lip positions were marked at three temporal points - onset, center, and offset - of the vowel in each test monosyllable for indicating the CS between the initial/final stop consonants and the neighboring vowels in the CV and VC contexts. The variations in the tongue and lip positions (*i*) between the vowel onset and vowel center pertaining to C-V co-articulation and (*ii*) between the vowel center and vowel offset pertaining to V-C co-articulation were also computed and expressed in terms of the Euclidean distance. A small or large Euclidean distance is taken to indicate a respective weak or strong CS between the neighboring consonants and vowels in the test syllables.

3 Results

Figures 1a–c, Figs. 3a–b, and Figs. 5a–c show the superimposed mid-sagittal tongue contours at the vowel onset (in red dotted lines) and vowel center (in blue solid lines) of [i a u] preceded by the initial stop [p-], [t-] or [k-] in Cantonese CV and CVC syllables for Male Speaker 1. Figures 2a–b, Figs. 4a–c, and Fig. 6 show the superimposed mid-sagittal tongue contours at the vowel center (in blue solid lines) and vowel offset (in green dashed lines) of [i a u] followed by the final stop [-p], [-t], or [-k] in Cantonese VC and CVC syllables also for Male Speaker 1. In the figures, the tongue contours were drawn by connecting the positions of TT, TM, and TB, with the front-back dimension on the x-axis against the up-down dimension on the y-axis. All the figures are on the same scale, with each grid side in 10 mm. The thick dark line in each figure is the tracing of the mid-sagittal palate contour of the speaker, facing to the left, serving as the reference, relative to which the tongue positions in both the front-back (x) and up-down (y) dimensions during the different temporal points of the vowels are defined and compared. The mid-sagittal tongue contours for Male Speaker 2 in this study are not presented, due to space limitation and the between-speaker similarities.

3.1 Bilabial Stop [p]

As shown in Figs. 1a–c, when the initial [p-] precedes a high vowel [i] (Fig. 1a) or [u] (Fig. 1c), the tongue positions at the vowel onset and vowel center are nearly the same, suggesting a strong CS between [p-] and the following high vowel. During the articulation of [p-], the tongue is not involved and is thus free to assume the anticipated gesture of the following vowel. For [p]-[i] (Fig. 1a), but not [p]-[u] (Fig. 1b), the position of TB tends to raise slightly closer to the palate at the vowel center than at the vowel onset, an indication that the tongue does not quite yet reach the target position at the vowel onset. The t-test results substantiate the finding, in that the differences in the positions of the three tongue points (TT, TM, TB) between the vowel onset and vowel center are non-significant for [p]-[i] or [p]-[u], except for the position of TB which is significantly higher at the vowel center than the vowel onset ($p < 0.001$) for [p]-[i].

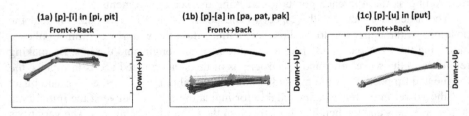

Fig. 1. a–c. Superimposed mid-sagittal tongue contours at the vowel onset (in red dotted lines) and vowel center (in blue solid lines) of [i a u] preceded by [p-] in Cantonese CV and CVC syllables and the palate contour (in thick dark line) for Male Speaker 1. (Color figure online)

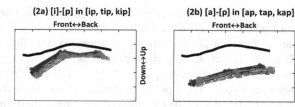

Fig. 2. a–b. Superimposed mid-sagittal tongue contours at the vowel center (in blue solid lines) and vowel offset (in green dashed lines) of [i a] followed by [-p] in Cantonese VC and CVC syllables and the palate contour (in thick dark line) for Male Speaker 1. (Color figure online)

As for [p]-[a] (Fig. 1b), differing from the cases of [p]-[i] (Fig. 1a) and [p]-[u] (Fig. 1c), the positions of all the three tongue points (TT, TM, TB) are higher at the vowel onset than at the vowel center and the difference is significant ($p < 0.001$), suggesting a decrease in CS between [p-] and [a]. With respect to the variations in lip position, the Euclidean distance between the two lips increases significantly at the vowel center as compared to the vowel onset when [p-] precedes [i], [u], or [a] ($p < 0.001$). The percentage of the increase in the Euclidean distance between the upper and lower lips at the vowel center is much larger for [p]-[a] (73%) than [p]-[i] (30%) and [p]-[u] (16%), attributed to the competition between the lip closure during [p-] and lip/mouth opening for articulating the open vowel [a]. The data on the variations in lip opening also indicate a weaker CS between [p-] and the open vowel [a] than between [p-] and the close vowel [i] or [u].

The CS is stronger between the final [-p] and the preceding vowel of any type in the VC context than between the initial [p-] and the corresponding vowel in the CV context. As shown in Fig. 2a and b, the tongue positions at the vowel center and vowel offset are similar when [-p] preceded by the close vowel [i] or the open vowel [a]. Nonetheless, for both [i]-[p] and [a]-[p], the positions of TT, TM, and TB are slightly retracted at the vowel offset as compared to the vowel center, and the differences are significant ($p < 0.001$). The data indicate that [-p] is not in total co-articulation with the preceding vowel, even though the tongue is not involved during the articulation of [-p].

With respect to the lip position, the Euclidean distance between the upper and lower lips is reduced at the vowel offset compared to the vowel center. The percentages of reduction in the distance for [i]-[p] (41%) and [a]-[p] (33%) are similar, which indicate comparable degrees of CS between the final [-p] and the preceding vowels of different types.

3.2 Alveolar Stop [t]

The CS is weak between the initial [t-] and the following vowel in the CV context, irrespective of the vowel type. As shown in Fig. 3a and b, the position of TT at the vowel onset is raised close to the anterior part of the palate compared to the lowered position of TT at the vowel center, and the difference in position is significant ($p < 0.001$). This is attributed to the antagonism between the articulations of [t-] and the neighboring vowel, where TT is raised to form the alveolar closure during [t-], but lowered for the articulation of the vowel. When [t-] precedes the close vowel [i] (Fig. 3a), the positions of TM and TB are raised close to the palate at both the vowel onset and vowel center. When [t-] precedes the open vowel [a] (Fig. 3b), the positions of TT, TM, and TB are higher at the vowel onset than at the vowel center, and the differences in position are all significant ($p < 0.001$), though the difference in the position of TB between the two temporal points is small. The data indicate that the CS is stronger for [t]-[i] than [t]-[a], due to the elevation of the tongue tip during [t-] which is more compatible with the tongue gesture for the articulation of the close vowel [i] than for the open vowel [a].

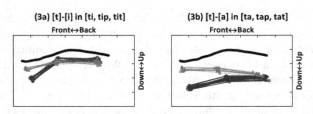

Fig. 3. a–b. Superimposed mid-sagittal tongue contours at the vowel onset (in red dotted lines) and vowel center (in blue solid lines) of [i a] preceded by [t-] in Cantonese CV and CVC syllables and the palate contour (in thick dark line) for Male Speaker 1. (Color figure online)

In the VC context, the CS between V and [-t] is weaker when V = [u] than when V = [i] or [a]. As shown in Fig. 4c, for [u]-[t], TT at the vowel offset is moved towards the anterior part of the palate, but not at the vowel center. The positions of TM and TB also differ between the two temporal points, being more fronted at the vowel offset than at the vowel center. The differences in the positions of the three tongue points between the vowel onset and vowel center are all significant ($p < 0.001$), indicating a weak CS between [u] and [-t]. As for [i]-[t] (Fig. 4a) and [a]-[t] (Fig. 4b), the raising of TT is not observed at the vowel offset, suggesting the absence of the alveolar closure for the final stop [-t]. It is assumed that [-t] is articulated as a glottal stop [-ʔ] instead. In general, when [i] or [a] followed by [-t], the tongue positions at the vowel center and vowel offset are similar, an indication of a strong CS between [i] or [a] and [-t].

3.3 Velar Stop [k]

There is a strong CS between the initial [k-] and the following high front vowel [i], attributed to the similarity in tongue position between [k-] and [i] as a result of the palatalization of [k-] before [i]. As shown in Fig. 5a, when [k-] precedes [i], the positions of TT, TM, and TB at the vowel onset are nearly the same as those at the vowel center. Statistical data show that the difference in tongue position between the two temporal points is mildly significant ($p < 0.05$) only on the y-coordinate plane. Relative to the case of [k]-[i] (Fig. 5a), the difference in tongue position between the vowel onset and vowel center increases when [k-] precedes a high back [u] (Fig. 5c) and further increases when [k-] precedes an open [a] (Fig. 5b), suggesting a decrease in

(4a) [i]-[t] in [it, pit, tit, kit]

Front↔Back

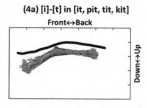

(4b) [a]-[t] in [at, pat, tat]

Front↔Back

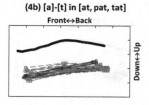

(4c) [u]-[t] in [ut, put]

Front↔Back

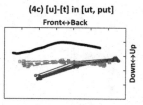

Fig. 4. a–c. Superimposed mid-sagittal tongue contours at the vowel center (in blue solid lines) and vowel offset (in green dashed lines) of [i a u] followed by [-t] in Cantonese VC and CVC syllables and the palate contour (in thick dark line) for Male Speaker 1. (Color figure online)

CS between [k-] and [u] or [a]. The difference in tongue position between the two temporal points is significant for [k]-[u] ($p < 0.01$) and in particular for [k]-[a] ($p < 0.001$) due to the antagonism between the tongue positions of [k-] and [a]. The tongue is raised towards the palate during [k-], but lowered during [a].

(5a) [k]-[i] in [ki, kip, kit]

Front↔Back

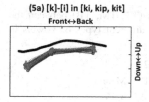

(5b) [k]-[a] in [ka, kap, kak]

Front↔Back

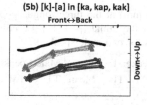

(5c) [k]-[u] in [ku]

Front↔Back

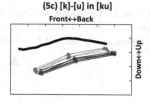

Fig. 5. a–c. Superimposed mid-sagittal tongue contours at the vowel onset (in red dotted lines) and vowel center (in blue solid lines) of [i a u] preceded by [k-] in Cantonese CV and CVC syllables and the palate contour (in thick dark line) for Male Speaker 1. (Color figure online)

In the VC context, the CS between the open vowel [a] and the final [-k] is strong. As shown in Fig. 6, there is a close similarity in tongue position between the vowel center and vowel offset for the single case of [a]-[k]. Statistical data show the differences in the positions of different tongue points, except for TB, are non-significant between the two temporal points. At the vowel offset, the positions of TT, TM, and TB are distant from the palate, suggesting an absence of the velar closure for the final stop [-k]. It is assumed that [-k] is articulated as a glottal stop [-ʔ] instead.

3.4 Between-Segment CS in CV and VC Contexts

This section presents the data on the between-segment CS expressed in terms of the Euclidean distance (*i*) between the tongue positions at the vowel onset and vowel center for C-V co-articulation and (*ii*) between the tongue positions at the vowel center and vowel offset for V-C co-articulation during the Cantonese monosyllables for both Male Speaker 1 and Male Speaker 2. Table 2 presents the data on the Euclidean distances (in mm) between the vowel onset and vowel center at the three tongue points (TT, TM, TB) and the mean values averaged across the three tongue points for the

(6) [a]-[k] in [ak, pak, kak]

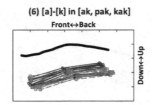

Fig. 6. Superimposed mid-sagittal tongue contours at the vowel center (in blue solid lines) and vowel offset (in green dashed lines) of [a] followed by [-k] in Cantonese VC and CVC syllables and the palate contour (in thick dark line) for Male Speaker 1. (Color figure online)

initial [p- t- k-] preceding different types of vowels in CV and CVC syllables. Table 3 presents the corresponding data on the Euclidean distances between the vowel center and vowel offset for the final [-p -t -k] following different types of vowels in VC and CVC syllables. A large or small Euclidean distance is taken to indicate a respective weak or strong CS between the stops and the neighboring vowels in the syllables.

Table 2. Euclidean distances (in mm) at the three tongue points (TT, TM, TB) between the vowel onset and vowel center for the initial [p- t- k-] preceding different types of vowels in Cantonese CV and CVC syllables for two male speakers.

	Male Speaker 1				Male Speaker 2			
	TT	TM	TB	Mean	TT	TM	TB	Mean
[p]-V	3.63	4.59	3.72	3.98	4.36	3.29	2.42	3.36
[t]-V	10.52	4.70	3.51	6.24	13.45	8.56	5.40	9.14
[k]-V	4.29	6.27	6.36	5.64	2.80	5.45	6.36	4.87

The data presented in Table 2 for the two Cantonese speakers show that the Euclidean distance at TB is larger for [k]-V (6.36/6.36 mm) than [p]-V (3.72/2.42 mm) and [t]-V (3.51/5.40 mm), irrespective of the types of the following vowels, whereas the Euclidean distance at TT is significantly larger for [t]-V (10.52/13.45 mm) than [p]-V (3.63/4.36 mm) and [k]-V (4.29/2.80 mm). This is because TB is involved in the articulation of [k-], but not [p-] and [t-], competing for the use of TB with the following vowel. As for TT, it is raised to form the alveolar closure during [t-], which is incompatible with the lowering of TT during the following vowel, but not during [p-] or [k-]. Based on the Euclidean distance averaged across the three tongue points, the order of decreasing Euclidean distance is [t]-V > [k]-V > [p]-V for both Male Speaker 1 (6.24 mm > 5.64 mm > 3.98 mm) and Male Speaker 2 (9.14 mm > 4.87 mm > 3.36 mm). Thus, the order of decreasing CS between the initial C- and the following vowel in the CV context is when C- = [p-] > C- = [k-] > C- = [t-].

The Euclidean distances at all the three tongue points are reduced for V-[p], V-[t], and V-[k] (Table 3), relative to [p]-V, [t]-V, and [k]-V (Table 2), indicating an increase in CS between the final stop and the preceding vowel in the VC context. Table 3 shows that for both Male Speakers 1 and 2, the Euclidean distance at TT is larger for V-[t]

Table 3. Euclidean distances (in mm) at the three tongue points (TT, TM, TB) between the vowel center and vowel offset for the final [-p -t -k] following different types of vowels in Cantonese VC and CVC syllables for two male speakers.

	Male Speaker 1				Male Speaker 2			
	TT	TM	TB	Mean	TT	TM	TB	Mean
V-[p]	3.32	2.14	1.55	2.34	3.39	1.94	1.94	2.42
V-[t]	4.71	3.33	2.83	3.62	10.14	4.86	3.69	6.23
V-[k]	2.32	2.72	2.97	2.67	2.62	2.18	2.62	2.47

(4.71/10.14 mm) than V-[p] (3.32/3.39 mm) and V-[k] (2.32/2.62 mm), which parallels the pattern of a larger Euclidean distance at TT for [t]-V than [p]-V and [k]-V in the CV context (Table 2).

With respect to the position of TB, the Euclidean distance is larger for V-[k] (2.97/2.62 mm) than V-[p] (1.55/1.94 mm) irrespective of the types of the preceding vowels for both Cantonese speakers, which also parallels the pattern of a larger Euclidean distance at TB for [k]-V (6.36/6.36 mm) than [p]-V (3.72/2.42 mm) in the CV context. As for the difference in the Euclidean distance at TB between V-[k] and V-[t], the patterns differ between the two speakers, in that the Euclidean distance at TB is slightly larger for V-[k] (2.97 mm) than V-[t] (2.83 mm) for Male Speaker 1, but smaller for V-[k] (2.62 mm) than V-[t] (3.69 mm) for Male Speaker 2. In the CV context, the Euclidean distance at TB is larger for [k]-V (6.36/6.36 mm) than [t]-V (3.51/5.40 mm) for both speakers.

Based on the Euclidean distance averaged across the three tongue points for each of three final stops [-p -t -k] preceded by different vowel types in the VC context, the order of decreasing Euclidean distance is V-[t] > V-[k] > V-[p] for both Male Speaker 1 (3.62 mm > 2.67 mm > 2.34 mm) and Male Speaker 2 (6.23 mm > 2.47 mm > 2.42 mm), though the difference in the Euclidean distance between V-[k] and V-[p] is only small. In general, based on the data on Euclidean distance, the order of decreasing CS for V-C co-articulation is when -C = [-p] > -C = [-k] > -C = [-t], which parallels the order of decreasing CS for C-V co-articulation, i.e., when C- = [p-] > C- = [k-] > C- = [t-].

4 Conclusion

This paper has presented the co-articulation strength between the initial/final stops [p- t- k-]/[-p -t -k] and vowels [i a u] in Cantonese monosyllables. The findings suggest the factors affecting the between-segment CS include the stop consonant type, vowel type, and initial/final position of the stop in the syllable. For both the C-V and V-C co-articulation, the order of decreasing CS is when C = [p] > C = [k] > C = [t], and the degree of CS is higher in the VC than the CV context. The articulatory data support the phonological structuring of the syllable, wherein the syllable-final consonant (or syllable coda), but not the syllable-initial consonant (or syllable onset), and the preceding vowel (or syllable nucleus) form the phonological unit of the rhyme. Also, the findings of the present study are generally in agreement with the patterns of co-articulation between consonant and vowel in other languages reported in previous studies.

Acknowledgement. This research is supported by an International Scholarly Exchange Research Grants (#9500023/RG003-P-14) from the Chiang Ching-kuo Foundation, Taiwan.

References

1. Duez, D.: Second formant locus-nucleus patterns: an investigation of spontaneous French speech. Speech Commun. **11**, 417–427 (1992)
2. Farnetani, E.: V-C-V lingual co-articulation and its spatiotemporal domain. In: Hardcastle, W.J., Marchal, A. (eds.) Speech Production and Speech Modelling, Ch. 5, pp. 93–130. Kluwer Academic Publishers (1990)
3. Fowler, C.A., Brancazio, L.: Co-articulation resistance of American English consonants and its effects on transconsonantal vowel-to-vowel co-articulation. Lang. Speech **43**, 1–41 (2000)
4. Iskarous, K., Fowler, C.A., Whalen, D.H.: Locus equations are an acoustic expression of articulator synergy. J. Acoust. Soc. Am. **128**, 2021–2032 (2010)
5. Krull, D.: Second formant locus patterns as a measure of consonant-vowel co-articulation. PERILUS (Phonetic Experimental Research at the Institute of Linguistics University of Stockholm) V, pp. 43–61 (1987)
6. Krull, D.: Consonant-vowel co-articulation in spontaneous speech and in reference words. (Doctoral dissertation). PERILUS (Phonetic Experimental Research at the Institute of Linguistics University of Stockholm) VII, pp. 1–149 (1988)
7. Krull, D.: Second formant locus patterns and consonant-vowel co-articulation in spontaneous speech. PERILUS (Phonetic Experimental Research at the Institute of Linguistics University of Stockholm), X, pp. 87–108 (1989)
8. Lindblom, B.: Spectrographic study of vowel reduction. J. Acoust. Soc. Am. **35**, 1773–1781 (1963)
9. Modarresi, G., Sussman, H.M., Lindblom, B., Burlingame, E.: Stop place coding: an acoustic study of CV, VC#, and C#V sequences. Phonetica **61**, 2–21 (2004)
10. Stevens, K.N., House, A.S.: Perturbation of vowel articulations by consonantal context: an acoustical study. J. Speech Hear. Res. **6**, 111–128 (1963)
11. Sussman, H.M., Bessell, N., Dalston, E., Majors, T.: An investigation of stop place of articulation as a function of syllable position: a locus equation perspective. J. Acoust. Soc. Am. **101**, 2826–2838 (1997)
12. Sussman, H.M., Hoemeke, K.A., Ahmed, F.S.: A cross-linguistic investigation of locus equations as a phonetic descriptor for place of articulation. J. Acoust. Soc. Am. **94**, 1256–1268 (1993)
13. Sussman, H.M., McCaffrey, H.A., Matthews, S.A.: An investigation of locus equations as a source of relational invariance for stop place categorization. J. Acoust. Soc. Am. **90**, 1309–1325 (1991)
14. Recasens, D.: V-to-C co-articulation in Catalan VCV sequences: an articulatory and acoustical study. J. Phonetics **12**, 61–73 (1984)
15. Recasens, D.: Coarticulatory patterns and degrees of coarticulatory resistance in Catalan CV sequences. Lang. Speech **28**, 97–114 (1985)
16. Recasens, D., Espinosa, A.: An articulatory investigation of lingual coarticulatory resistance and aggressiveness for consonants and vowels in Catalan. J. Acoust. Soc. Am. **125**, 2288–2298 (2009)

Speech Style Effects on Local and Non-local Coarticulation in French

Giuseppina Turco[2], Fanny Guitard-Ivent[1], and Cécile Fougeron[1(✉)]

[1] Laboratoire de Phonétique et Phonologie, UMR 7018,
CNRS/Sorbonne Nouvelle, Paris, France
{fanny.ivent,cecile.fougeron}@sorbonne-nouvelle.fr
[2] Laboratoire de Linguistique Formelle, UMR 7110, CNRS/Paris-Diderot,
Paris, France
gturco@linguist.univ-paris-diderot.fr

Abstract. Interactions between speech style and coarticulation are investigated by examining local, non-local, anticipatory and carryover contextual effects on vowels in two French corpora of conversational and journalistic speech. C-to-V coarticulation is analyzed on 22 k tokens of /i, E, a, u, ɔ/(/E/=/e, ɛ/) 50-to-80 ms long. Contextual effects are measured as F2 changes in relation to the adjacent consonant (alveolar *vs.* uvular) in CV_1 and V_1C sequences. V-to-V coarticulation is analyzed on 33 k $V_1C(C)V_2$ sequences with V_1 = /e, ɛ, o, ɔ, a/ falling within the same range of duration, and V_2 either high/mid-high or low/mid-low. Contextual effects are measured as F1 changes as function of V_2 height. Results show more local C-to-V coarticulation in conversational than in journalistic speech, as previously found for other languages. Interestingly, this interaction is clearer for all vowels in V_1C, whereas coarticulation in CV_1 is affected by style for non-high vowels only. V-to-V coarticulation is also found in both corpora but is modulated by style only for mid-front vowels and in the opposite direction (i.e. more overlap in journalistic than in conversational speech). Findings are interpreted in light of dynamic models of speech production and of a phonological account of French V-to-V harmony.

Keywords: Local *vs.* non-local coarticulation · Speech style · French
Large corpora · Acoustic · Anticipatory *vs.* carryover coarticulation

1 Introduction

Pronunciation is highly sensitive to speech style variations. In the literature, reported acoustic changes concern the spectral and temporal properties of a sound. Specifically, previous studies have shown that segment lengthening (Ferguson and Kewley-Port 2002, 2007; Picheny *et al.* 1986; Moon and Lindblom 1994) and hyperarticulation, often measured with an expansion of the vowel acoustic space (Bradlow 2002; Johnson *et al.* 1993; Picheny *et al.* 1986; Audibert *et al.* 2015; Gendrot and Adda-Decker 2005), represent robust cues of "clear" (e.g. overarticulated read speech) speech compared to other forms of "elicited" speech (e.g. normal read speech) as well as "casual" speech (often referred to as spontaneous speech).

© Springer Nature Switzerland AG 2018
Q. Fang et al. (Eds.): ISSP 2017, LNAI 10733, pp. 121–133, 2018.
https://doi.org/10.1007/978-3-030-00126-1_12

Less understood is the relation between speech style and coarticulation. Some studies revealed a reduction of coarticulation in clear speech (Krull 1989 for Swedish; Duez 1992 for French; Moon and Lindblom 1994 for English; DiCanio et al. 2015 for Mixtec) compared to less clear speech (citation-form or casual-form). Other studies did not find any adjustments in degree of coarticulation according to speech style (Matthies et al. 2001; Bradlow 2002). A third type of (more perception-oriented) study (Scarbourough and Zellou 2013) found that it is the authenticity of the communicative context to induce variation in coarticulation degree (i.e. more coarticulation in a "real" than in a "simulated" clear speech situation).

These different findings may be due to several reasons. First of all, they test different directions of coarticulation reflecting two distinct underlying processes: anticipatory coarticulation (as in Krull 1989; Duez 1992, Scarborough and Zellou 2013) vs. carryover coarticulation (as in Moon and Lindblom 1994; Bradlow 2002; DiCanio et al. 2015). While both types of coarticulation can result from gestural overlap, anticipatory coarticulation can also reflect the planning of upcoming speech units, while carryover coarticulation can result from mechanical or inertial effects of moving an articulator from one target to the next one (e.g. Recasens 1987). Second, it is questionable whether the elicited speech conditions tested in these studies (from clear speech to more spontaneous speech) are comparable. Finally, there are cross-linguistic differences in coarticulatory patterns (Öhman 1966; Manuel 1990; Beddor 2002; Ma et al. 2006) that may reflect differences in the relation between style and coarticulation across languages.

The current study investigates potential effects of speech style on vowel coarticulation in two big French corpora of (natural) casual conversation and of (natural) formal journalistic speech. Different from previous work mostly carried out on elicited speech, we investigate the relation between speech style and coarticulation in a more ecologically-valid speaking condition.

The novel aspect of the investigation lies in the comparison between CV_1 sequences with potential carryover coarticulation, V_1C sequences with anticipatory coarticulation, and $V_1C(C)V_2$ sequences with V-to-V anticipatory coarticulation. Hence, with respect to prior research mainly focusing on contextual overlap operating at a local level, here the observed coarticulation varies in terms of *distance* (local vs. non-local) and *cohesion* between the segments in context. The inter-articulator coordination within CV_1 sequences (coupled in-phase) has proven to be more stable than in (anti-phase) V_1C sequences (Kelso et al. 1986, Tuller and Kelso 1991, Nam et al. 2010). In addition, compared to C-to-V, V-to-V is the least cohesive structure, in which coordination seems to be the most language-dependent.

Another novel aspect of our study is the comparison among different sets of vowels. C-to-V coarticulation is analyzed on tokens of /i, E, a, u, ɔ/ (with /E/=/e, ɛ/). These are the vowels for which we had enough material available in our two corpora in V_1C and CV_1 sequences. Furthermore, the selection of this set of vowels allows us to test at the same time back vs. front vowels as well as high vs. non-high vowels. We expect the degree of coarticulation between C and V be modulated by speech style. At the same time, we set to determine whether the relation between coarticulation and style depends on the type of vowel. According to the Degree of Articulatory Constraints (DAC) model, vowels articulated with a palatal constriction are said to be more constrained articulatorily, and therefore, more resistant to contextual variation than others (Recasens 2007).

V-to-V coarticulation (in $V_1C(C)V_2$ sequences) is tested on V_1 containing mid-front and mid-back vowels (=/e, ɛ, o, ɔ/) as well as the low vowel (=/a/). While low vowels are particularly noteworthy for their high degree of contextual variation (e.g., Recasens 1987), mid-front and mid-back vowels are known to undergo a phonological process of anticipatory vowel harmony in French, at least for the Northern varieties (which are closer to Standard French, see Nguyen and Fagyal 2008; Turco et al. 2016). The target V_1 usually alternates in height (mid-high/mid-low) according to the height changes of the following V_2 (high vs. low, respectively). The classic literature on French vowel harmony (e.g., Fouché 1959, Walker 2001) suggests that the degree of influence of V_2 on V_1 may be sensitive to speech style variations. V_1 harmonizes more with V_2 in casual than in formal speech, an observation that is largely consistent with previous findings on local coarticulation and speech style. With respect to previous work, here we aim at investigating the relation between vowel harmony and style more systematically (see also Turco *et al.* 2016 for a preliminary investigation).

A further open question in the literature is whether variations affecting the spatial (formant) properties of the vowel are induced by variations in duration. In a revised version of Lindblom (1963)'s vowel *undershoot* model, Moon and Lindblom (1994) showed that vowel formant displacements due to consonantal context highly depended on vowel duration and that the link between spectral changes and duration was less strong in clear-speech than in citation-form speech (a similar finding is also reported by Duez 1992 for V-to-C coarticulation in French). However, we know that speech style variation is more than just a matter of duration: speakers can adopt different production strategies according to the speaking condition at hand (e.g. Ferguson and Kewley-Port 2002; Smiljanic and Bradlow 2005). As a matter of fact, previous work on speech style differences in French has shown that vowel spectral reduction in conversational speech compared to read speech mainly applied when looking at short vowels (under 40 ms; see, for instance, Rouas *et al.* 2010). Moreover, Audibert *et al.* (2015) found more reduction of the acoustic space, more vowel centralization and intra-category dispersion in conversational speech than in journalistic and read speech for short vowels (up to 50 ms) than for mid (up to 80 ms) and long vowels (up to 120 ms). Hence, on the basis of these findings, we test potential effects of speech style on coarticulation by controlling for the duration of the target vowel.

To summarize, the questions we address in the current study are the following ones:

1. whether degree of coarticulation varies with speech style (and, if so, in which direction and for which vowels);
2. whether speech style variation equally affects all types of coarticulation (anticipatory vs. carryover, local vs. non-local).
3. whether speech style variation affects vowel coarticulation when its duration is controlled for.

2 A Corpus-Based Study

Speech material was extracted from two publicly available French corpora: ESTER (Gravier *et al.* 2006) and NCCFr (Torreira *et al.* 2010). ESTER contains broadcasted news and political/societal debates of several radio and TV programs. It includes mainly scripted speech mostly produced by professional speakers. NCCFr contains conversational speech based on free and guided face-to-face discussions on societal topics among pair of friends (mainly young students). In our study, the corpus ESTER will be referred to as "journalistic" speech (J, henceforth), the corpus NCCFr as "conversational" speech (C).

2.1 Speech Material and Data Coding

Local C-to-V coarticulation was observed on 22 k tokens extracted from both corpora. The target vowel (V_1) was represented by the following set of sounds: /i, E, a, u, ɔ/ (with /E/=/e, ɛ/), produced by 23 male speakers. The duration of the target vowel was controlled: all vowels ranged from 50 to 80 ms. The adjacent consonant was either an alveolar (C_{ALV} =/t, d, z, s, l, n/, e.g. *dépanner* /depane/- "to help"), which is known to attract F2 towards an 1800 Hz locus, or a uvular (C_{UV} =/R/, e.g. *appareil* /apaRɛj /- "appliance"), known to lower F2 (and raise F1). The opposite context (left in V_1C and right in CV_1) was always a labial consonant. Note that in CV_1 sequence, C and V_1 are always tautosyllabic, while in V_1C sequences syllabification could not be controlled for. Contextual effects were measured as changes of the second formant (F2) of V_1 according to the place of articulation of the adjacent C.

V-to-V coarticulation was analyzed on 33 k words containing a $V_1C(C)V_2$ sequence where the target V_1 was a mid-high, mid-low or low vowel (/e, ɛ, o, ɔ, a/) located in the penultimate syllable of the word. The influencing vowel V_2 was either a raising trigger (high and mid-high /i, e, o, y, u/, all coded as 'high') or a lowering trigger (low and mid-low /ɛ, a, ã/, all coded as 'low') placed in the last (accented) syllable. In the extraction of those words, no constraint on their syllable structure and their sequence of intervocalic consonants (i.e. between V_1 and V_2) was applied. The degree of coarticulation was measured as the lowering of the first formant (F1) of V_1 in relation to the height of V_2 (high/mid-high vs. low/mid-low).

The vowels were identified according to a forced automatic alignment (using the Burg algorithm in the Praat software, Boersma and Weenink 2018) that takes into account both orthographic and phonological information at a word level (see Turco *et al.* 2016 for details). In all cases, the duration of V_1 was controlled to avoid duration-dependent variation in coarticulation degree. All vowels were between 50 ms and 80 ms. These duration values were obtained following classification criteria applied to the same corpora as described in Audibert *et al.* (2015). Finally, formant values were extracted at 1/3, 1/2 and 2/3 of the vowel and then averaged for a single value per vowel. In order to check for outliers (due, for instance, to mislabelling), formant values were inspected according to the procedure described in Gendrot and Adda-Dekker (2005). They were then *z*-score transformed (Lobanov 1971) by subtracting the mean of all data points from each of them and by dividing those points by the standard deviation of all data points.

2.2   Statistical Analyses

We performed linear mixed-effects models using the lme4 package (Bates *et al.* 2015) in the R software (R Development Core Team 2008).

Two types of model structure were constructed. For local C-to-V coarticulation, we modeled the relationship between the F2 of the target V_1 (in z-score) in relation to the adjacent CONSONANTAL CONTEXT (C: alveolar *vs.* uvular), DIRECTION TYPE (CV_1 *vs.* V_1C) and SPEECH STYLE (conversational *vs.* journalistic speech). For non-local V-to-V coarticulation, the model contained the F1 of V_1 (in z-score) as function of V_2 HEIGHT (high *vs.* low) and SPEECH STYLE (conversational *vs.* journalistic speech). In both models, speaker was included as a random factor. By-speaker random slopes were also included to avoid high Type I error rate (Cunnings 2012). The two models were then run separately for each vowel according to the type of coarticulation they belonged to (C-to-V or V-to-V). *P*-value estimates were based on *Satterthwaite* approximations, which provides more conservative estimates for linear regression, through the *lmerTest*()- function (Kuznetsova *et al.* 2013) and further adjusted for multiple testing comparisons. Likelihood ratio tests as implemented in the *anova*()-function were performed to check main effects of each fixed factor and interactions.

In line with previous studies (cf. Introduction), we predicted more coarticulation in conversational than in formal journalistic speech. In statistical terms, this is translated into an interaction between the relevant predictors (i.e. between CONSONANTAL CONTEXT and SPEECH STYLE for CV_1 and V_1C sequences; V_2 HEIGHT and SPEECH STYLE for V_1CCV_2 sequences). In what follows, we present only the relevant effects and interactions that reply to our research questions (cf. Introduction).

3   Results

3.1   C-to-V Coarticulation

First, the statistical analyses reveal a significant main effect of CONSONANTAL CONTEXT on the F2 of all the tested vowels (see Table 1 below). As expected, the F2 of the target vowels is higher when these vowels are in alveolar than in a uvular context.

Table 1. Estimates (β-coefficients, standard errors (SE), *t*-values and *p*-values) of the linear mixed effects models for the effect of CONSONANTAL CONTEXT on the F2 of five vowels (/i/, /u/, E =/ e, ɛ/, /ɔ/, /a/). The intercept contains "alveolar" as a reference value; "n.s." stands for *p*-values that are not significant.

CONSONANTAL CONTEXT			
	β_{uvular}	SE	t	*p*-value
/i/	−.245	.055	−4.42	.0006
/u/	−.390	.056	−6.91	.0001
/E/	−.090	.031	−2.88	.01
/ɔ/	−.380	.073	−5.17	.0001
/a/	−.471	.048	−9.90	.0001

By contrast, unexpectedly, an effect of DIRECTION TYPE is found for the vowels /E/, /ɔ/ and /a/: the F2 of these vowels is higher in CV_1 sequences than in V_1C sequences (see Table 2).

Table 2. Estimates (β-coefficients, standard errors (SE), t-values and p-values) of the linear mixed effects models for the effect of DIRECTION TYPE on the F2 of five vowels (/i/, /u/, E =/e, ɛ/, /ɔ/ , /a/). The intercept contains "VC" sequence as a reference value; "n.s." stands for p-values that are not significant.

DIRECTION TYPE				
	β_{CV}	SE	t	p-value
/i/	.017	.016	1.08	n.s.
/u/	.089	.061	1.44	n.s.
/E/	.254	.027	9.25	.001
/ɔ/	.263	.042	6.27	.0001
/a/	.148	.033	4.50	.0001

Furthermore, our analyses reveal an effect of SPEECH STYLE on F2 for certain vowels only (Table 3): their F2 is higher in journalistic speech than in conversational speech.

Table 3. Estimates (β-coefficients, standard errors (SE), t-values and p-values) of the linear mixed effects models for the effect of STYLE on the F2 of five vowels (/i/, /u/, E =/e, ɛ/, /ɔ/, /a/). The intercept contains "journalistic" as a reference value; "n.s." stands for p-values that are not significant.

STYLE				
	$\beta_{conversational}$	SE	t	p-value
/i/	.136	.048	2.83	.009
/u/	−.096	.089	−1.08	n.s.
/E/	.085	.034	2.48	.02
/ɔ/	.250	.086	2.90	.008
/a/	.008	.063	.13	n.s.

More interestingly, an interaction between CONSONANTAL CONTEXT and SPEECH STYLE is found for all the tested vowels: there is less coarticulation in journalistic speech than in conversational speech (Table 4).

Finally, the models reveal an interaction between CONSONANTAL CONTEXT, SPEECH STYLE and DIRECTION TYPE for the vowels /i, u, a, ɔ/ (see Table 5). Indeed, for close vowels /i, u/, the interaction is found in anticipatory V_1C coarticulation only (see Fig. 1a). For non-close vowels /ɔ, a/, this effect is present in both V_1C and CV_1 sequences. Yet, the reduction of coarticulation in journalistic speech is more important in V_1C than in CV_1 sequence (see Fig. 1b). In other words, SPEECH STYLE seems to affect mostly anticipatory coarticulation. For /E/, DIRECTION TYPE does not affect the interaction between CONSONANTAL CONTEXT and SPEECH STYLE: less coarticulation in journalistic speech is found in both V_1C and CV_1 sequence, as illustrated in Fig. 1c.

Table 4. Estimates (β-coefficients, standard errors (SE), *t*-values and *p*-values) of the linear mixed effects models for the interaction between CONSONANTAL CONTEXT*SPEECH STYLE on the F2 of five vowels (/i/, /u/, E =/e, ɛ/, /ɔ/, /a/). The intercept contains alveolar and journalistic as reference values; "n.s." stands for *p*-values that are not significant.

CONSONANTAL CONTEXT*SPEECH STYLE			
$B_{uvular*conversational}$	SE	*t*	*p*-value
/i/	−.265	.080	−3.31	.003
/u/	−.359	.077	−4.62	.0001
/E/	−.238	.041	−5.75	.0001
/ɔ/	−.714	.093	−7.67	.0001
/a/	−.344	.059	−5.80	.0001

Table 5. Estimates (β-coefficients, standard errors (SE), *t*-values and *p*-values) of the linear mixed effects models for the interaction between CONSONANTAL CONTEXT*DIRECTION TYPE*STYLE on the F2 of five vowels (/i/, /u/, E =/e, ɛ/, /ɔ/, /a/). The intercept contains alveolar, VC and journalistic as reference values; "n.s." stands for *p*-values that are not significant.

CONSONANTAL CONTEXT*DIRECTION TYPE*STYLE			
$B_{uvular*CV*conversational}$	SE	*t*	*p*-value
/i/	.216	.090	2.40	.03
/u/	.328	.143	2.30	.03
/E/	−.020	.065	−.30	*n.s.*
/ɔ/	.277	.111	2.49	.02
/a/	.187	.065	2.88	.009

3.2   V-to-V Coarticulation

For $V_1C(C)V_2$ sequences, the models reveal that V-to-V coarticulation is found for all the tested V_1 in the expected direction: F1 is lower when followed by a high V_2 (see Table 6 for model output). However, the change in F1 conditioned by V_2 HEIGHT is larger in magnitude for the mid-back vowels /o, ɔ/ and for the low /a/ than for the mid-front vowels /e, ɛ/.

The main effect of STYLE is found for the vowel /a/ only, showing a higher F1 in journalistic speech than in conversational speech (cf. Table 7). Together with the higher F2 for /i/ and the lower F2 for /u/ described in the previous section (cf. Sect. 3.1), the higher F1 of /a/ in journalistic speech contributes to the expansion of the vowel acoustic space expected in journalistic speech.

Crucially, the models reveal an interaction between V_2HEIGHT and STYLE for the E =/e, ɛ/ only (cf. Table 8). For the mid-front V_1, there is less coarticulation in conversational speech than in journalistic speech (see Fig. 2a). On the contrary, there is much more contextual change of F1 for the low and mid-back vowels, but this holds true for both speech styles (Fig. 2b and c), as indicated by the lack of a statistical interaction.

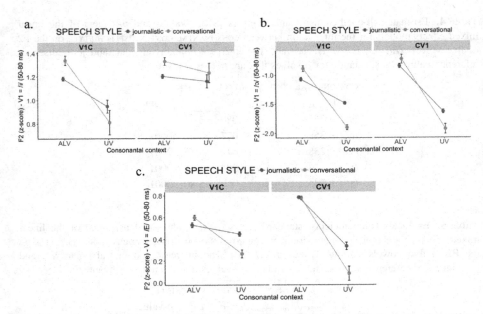

Fig. 1. F2 (in *z*-score) of the vowels /i/(panel a), /ɔ/(panel b) and /E/(panel c) ranging from 50 to 80 ms in V₁C and CV₁ context as function of CONSONANTAL CONTEXT (alveolar *vs*. uvular) and SPEECH STYLE (journalistic *vs*. conversational).

Table 6. Estimates (β-coefficients, standard errors (SE), *t*-values and *p*-values) of the linear mixed effects models for the effect of V₂ HEIGHT on the F1 of three vowels (/a/, E =/e, ɛ/, O =/o, ɔ/). The intercept contains "high" as a reference value; "n.s." stands for *p*-values that are not significant.

V₂ HEIGHT				
	β_low	SE	*t*	*p*-value
/a/	.069	.004	17.50	.0001
/E/	.032	.004	7.60	.0001
/O/	.124	.006	20.41	.0001

Table 7. Estimates (β-coefficients, standard errors (SE), *t*-values and *p*-values) of the linear mixed effects models for the effect of STYLE on the F1 of three vowels (/a/, E =/e, ɛ/, O =/o, ɔ/). The intercept contains "journalistic" as a reference value; "n.s." stands for *p*-values that are not significant.

STYLE				
	β_conversational	SE	*t*	*p*-value
/a/	−.041	.014	−2.90	.006
/E/	−.009	.014	−.63	*n.s.*
/O/	−.009	.016	−.56	*n.s.*

Table 8. Estimates (β-coefficients, standard errors (SE), *t*-values and *p*-values) of the linear mixed effects models for the interaction between V$_2$ HEIGHT*STYLE on the F1 of three vowels (/a/, E =/e, ɛ/, O =/o, ɔ/). The intercept contains "high" and "journalistic" as reference values; "n.s." stands for *p*-values that are not significant.

V$_2$ HEIGHT*STYLE				
	β$_{low*conversational}$	SE	*t*	*p*-value
/a/	.008	.006	1.29	*n.s.*
/E/	−.017	.006	−2.66	.01
/O/	.003	.011	.29	*n.s.*

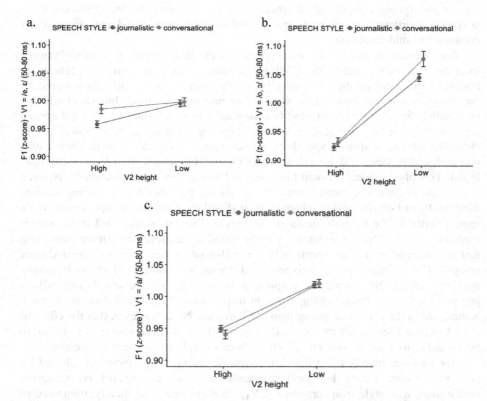

Fig. 2. F1 (in *z*-score) of the vowels (V$_1$) /e, ɛ/ (panel a), /o, ɔ/ (panel b) and /a/ (panel c) ranging from 50 to 80 ms as function of V$_2$ HEIGHT (high *vs.* low) and SPEECH STYLE (journalistic *vs.* conversational).

4   Discussion and Conclusion

Based on large French corpora of journalistic and conversational speech, in the current study we set to determine (1) whether the degree of segment overlap varies with speech style (and, if so, in which direction); (2) whether local carry-over CV$_1$, anticipatory

V_1C and non-local anticipatory $V_1C(C)V_2$ vary with style (i.e. whether the effect of style on the magnitude of coarticulation in the three cases tested here can tell us something about the difference between these three processes); (3) whether vowel coarticulation is affected by style when its duration is controlled for.

In line with previous studies (Krull 1989; Duez 1992; Moon and Lindblom 1994; DiCanio *et al.* 2015), for C-to-V coarticulation it was found that degree of overlap changed according to speech style – vowels were less coarticulated in formal (jour-nalistic) than in informal (conversational) speech; for V-to-V, the opposite pattern was observed – there was some more coarticulation in formal speech than in informal speech. This interaction occurred for the mid-front vowels only, that is, for that set of vowels undergoing a phonological process of vowel harmony. For mid-back and low / a/ vowels, coarticulation occurred regardless of style differences (and was even stronger compared to mid-front vowels).

The question is why are there differences in the direction of style effect between local and non-local coarticulation? While our findings are consistent with preliminary findings carried out on the same set of data (Turco et al. 2016), they do not confirm previous observations found in the classic literature for French (cf. Introduction). One explanation for why mid-front vowels assimilate less in informal than formal speech may be because we are dealing with a phonological phenomenon: vowel height changes affecting mid-front vowels in French may be considered as the reflex of a phonological process, and as such it shows crucial differences from V-to-V coarticu-lation. This phonological account finds also indirect evidence in the study by Nguyen and Fagyal (2008). The authors noticed that, during the laboratory recording session, their Southern French speakers changed their (Southern French) colloquial style into a more (Northern-like French) formal reading style. This "style-shift" led those partici-pants to produce more vowel harmony in the tested words. It is furthermore interesting that this interaction occurs specifically for mid-front vowels (and not for mid-back vowels). Considering that our conversational corpus is made of productions by young speakers (unlike the journalistic corpus), it is possible that we are dealing with a phenomenon of diachronic change. Height distinctions of mid-front vowels are more neutralized in the speech of young than old Parisians. Note, however, that the effect of the interaction (despite differences in the direction) was not as strong as it turns out to be for the C-to-V coarticulation, which leads us to reply to our second question.

Our analyses revealed that the three types of coarticulation were not affected by style to the same extent. For local coarticulation, (V_1C) anticipatory coarticulation varied more with style than carryover (CV_1) coarticulation. Specifically, compared to C_1V, V_1C showed less coarticulation in formal (journalistic) speech than in informal (conversational) speech for the set of vowels /i, u, O, a/, whereas for /E/ the degree of overlap modulated by style was the same across the two types of sequences. Moreover, compared to local coarticulation, there was less degree of overlap varying with style in non-local V-to-V coarticulation (V_1CCV_2). So, the crucial question here is why CV_1 and V_1C should vary – though differently – with style whereas V-to-V does not.

For local coarticulation, the 'coupled oscillator' model seems to best account for the differences observed between V_1C and CV_1 sequences. Being characterized by a stable inter-articulator timing coordination, it is conceivable that a CV_1 sequence (coupled in-phase) shows less sensitivity to speech style variation. By contrast, based on an anti-phase coupling, in a V_1C sequence the coordination of segments is less stable and hence may be more subject to potential sources of variation such as style (Kelso et al. 1986; Tuller and Kelso 1991, Nam et al. 2010). On the other hand, non-local V-to-V coarticulation (in extent, direction and magnitude) is language-dependent (Manuel 1990, 1999), speaker-dependent (Magen 1997) and, more importantly, dependent on the segment in the sequence over which is planned (Whalen 1990). Variability in V-to-V coarticulation has been shown to be mainly due to the articulatory demands of the intervening consonant (Recasens 1987; Ohman 1966), that is, to mechanical restrictions on tongue movements (e.g. degree of tongue dorsum contact). Crucially, these mechanical restrictions have been found to be more important for V-to-V anticipatory coarticulation than for carryover coarticulation (Recasens 1987) and seem to not vary with style. In Turco *et al.* (2016), more transconsonantal coarticulation was indeed found with a less constraining consonant (i.e. labials, as also found by Recasens 1987 on V-to-V in Catalan) between the two mid-vowels (i.e. vowel harmony). By contrast, in the current study based on the same corpora as Turco *et al.* (2016), a much larger variety of pivot consonants were included between the two vowels. It is hence likely that V-to-V coarticulation may remain unaffected by style variation since the planning mechanism (i.e. anticipating the second trigger vowel during the first target vowel) on which V-to-V is based, depends essentially on the nature of the intervening consonant, and not on style or other factors like speech rate and stress (Recasens 2015).

Finally, our findings suggest that the effect of speech style on coarticulation is not a by-product of durational variation. In line with previous research on the effect of style on the acoustics of vowels with controlled duration (Rouas *et al.* 2010 and Audibert *et al.* 2015); we have been able to show effects of style on degree of coarticulation of vowels falling within a certain range of duration (50–80 ms). In light of these findings, it therefore seems important to consider style as a factor of variation in its own right. Future studies should test if similar or different effects apply to other categories of vowel duration and if so, their implications for speech production models.

Acknowledgements. This research was funded by a research program and innovation of the European Union Horizon 2020, through a Marie Sklodowska-Curie n° 661530 awarded to the first author and by the ANR-10-LABX-0083 (Labex EFL).

References

Audibert, N., Fougeron, C., Gendrot, C., Adda-Decker, M.: Duration - vs. style-dependent variation: a multiparametric investigation. In: Proceedings of the 2015 ICPhS Conference, Glasgow (2015). (https://www.internationalphoneticassociation.org/icphs-proceedings/ICPhS2015/Papers/ICPHS0753.pdf)

Bates, D., Maechler, M., Bolker, B., Walker, S.: Fitting linear mixed-effects models using lme4. J. Stat. Softw. **67**(1), 1–48 (2015) https://doi.org/10.18637/jss.v067.i01

Beddor, P.S., Harnsberger, J.D., Lindemann, S.: Language-specific patterns of vowel-to-vowel coarticulation: acoustic structures and their perceptual correlates. J. Phonetics **30**(4), 591–627 (2002)

Boersma, P., Weenink, D.: Praat: doing phonetics by computer [Computer program]. Version 6.0.40 (2018). http://www.praat.org/. Accessed 11 May 2018

Bradlow, A.R.: Confluent talker-and listener-oriented forces in clear speech production. Lab. Phonology **7**, 241–272 (2002)

Cunnings, I.: An overview of mixed-effects statistical models for second language researchers. Second Lang. Res. **28**(3), 369–382 (2012)

Dicanio, C., Nam, H., Amith, J.D., García, R.C., Whalen, D.H.: Vowel variability in elicited versus spontaneous speech: evidence from Mixtec. J. Phonetics **48**, 45–59 (2015)

Duez, D.: Second formant locus-nucleus patterns: an investigation of spontaneouos French speech. Speech Commun. **11**(4–5), 417–427 (1992)

Ferguson, S.H., Kewley-Port, D.: Vowel intelligibility in clear and conversational speech for normal-hearing and hearing-impaired listeners. J. Acoust. Soc. Am. **112**(1), 259–271 (2002)

Ferguson, S.H., Kewley-Port, D.: Talker differences in clear and conversational speech: acoustic characteristics of vowels. J. Speech Lang. Hear. Res. **50**, 1241–1255 (2007)

Fouché, P.: Traité de prononciation française. Klincksieck, Paris (1959)

Gendrot, C., Adda, M.: Impact of duration on F1/F2 formant values of oral vowels: an automatic analysis of large broadcast news corpora in French and German. In: Proceedings of Eurospeech. Portugal, Lisbon, pp. 2453–2456 (2005)

Gravier, G., Bonastre, J.-F., Geoffrois, E., Galliano, S., Mc Tait, K., Choukri, K.: Corpus description of the ESTER evaluation campaign for the rich transcription of French broadcast news. In: Proceedings of European Conference on Speech Communication and Technology, pp. 139–142 (2006)

Johnson, K., Flemming, E., Wright, R.: The hyperspace effect: phonetic targets are hyperarticulated. Language **69**, 505–528 (1993)

Kelso, J.A., Saltzman, E.L., Tuller, B.: The dynamical perspective on speech production: data and theory. J. Phonetics **14**(1), 29–59 (1986)

Krull, D.: Second formant locus patterns and consonant-vowel coarticulation in spontaneous speech. Perilus **10**, 87–108 (1989)

Kuznetsova, A., Brockhoff, P.B., Christensen, R.H.B.: Lmertest: tests for random and fixed effects for linear mixed effect models (lmer objects of lme4 package). R package version 2. Pp. 0–29 (2013)

Lindblom, L.: On vowel reduction. The Royal Institute of Technology, Speech Transmission Laboratory, 29 (1963)

Lobanov, B.M.: Classification of Russian vowels spoken by different speakers. J. Acoust. Soc. Am. **49**, 606–608 (1971)

Ma, L., Perrier, P., Dang, J.: Anticipatory coarticulation in vowel-consonant-vowel sequences: a crosslinguistic study of French and Mandarin speakers. In: Proceedings of the 7th International Seminar on Speech Production, Ubatuba. pp. 151–158 (2006)

Magen, H.S.: The extent of vowel-to-vowel coarticulation in English. J. Phon. **25**(2), 187–205 (1997)

Manuel, S.Y.: The role of contrast in limiting vowel-to-vowel coarticulation in different languages. J. Acoust. Soc. Am. **88**(3), 1286–1298 (1990)

Manuel, S.: Cross-language studies: relating language-particular coarticulation patterns to other language-particular facts. In: Coarticulation: Theory, data and techniques, pp. 179–198 (1999)

Matthies, M., Perrier, P., Perkell, J.S., Zandipour, M.: Variation in anticipatory coarticulation with changes in clarity and rate. J. Speech Lang. Hear. Res. **44**, 340–353 (2001)

Moon, S.J., Lindblom, B.: Interaction between duration, context, and speaking style in English stressed vowels. The Journal of the Acoustical society of America **96**(1), 40–55 (1994)

Nam, H., Goldstein, L., Saltzman, E.: Self-organization of syllable structure: A coupled oscillator model. In: Pellegrino, F. (ed.) Approaches to phonological complexity, vol. 16, pp. 299–328. Walter de Gruyter, Berlin (2010)

Nguyen, N., Fagyal, Z.: Acoustic aspects of vowel harmony in French. J. Phon. **36**(1), 1–27 (2008)

Öhman, S.E.: Coarticulation in VCV utterances: Spectrographic measurements. J. Acoust. Soc. Am. **39**(1), 151–168 (1966)

Picheny, M.A., Durlach, N.I., Braida, L.D.: Speaking clearly for the hard of hearing. II: Acoustic characteristics of clear and conversational speech. J. Speech Hear. Res. **29**, 434–446 (1986)

R Development Core Team: A language and environment for statistical computing. R Foundation for Statistical Computing: Vienna (2008). www.r-project.org

Recasens, D.: An acoustic analysis of V-to-C and V-to-V: coarticulatory effects in Catalan and Spanish VCV sequences. J. Phonetics **15**(1), 71–86 (1987)

Recasens, D.: Patterns of VCV coarticulatory direction according to the DAC model. In: Prieto, P., Mascaró, J., Solé, M.J. (eds.) Segmental and Prosodic Issues in Romance Phonology, pp. 25–40. John Benjamins, Amsterdam (2007)

Recasens, D.: The effect of stress and speech rate on vowel coarticulation in catalan vowel–consonant–vowel sequences. J. Speech Lang. Hear. Res. **58**(5), 1407–1424 (2015)

Rouas, J.L., Beppu, M., Adda-Decker, M.: Comparison of spectral properties of read, prepared and casual speech in French. In: International Conference on Language Resources and Evaluation (LREC 2010), Malta, pp. 606–611 (2010)

Scarborough, R., Zellou, G.: Clarity in communication: "Clear" speech authenticity and lexical neighborhood density effects in speech production and perception. J. Acoust. Soc. Am. **134** (5), 3793–3807 (2013)

Smiljanić, R., Bradlow, A.R.: Production and perception of clear speech in Croatian and English a. J. Acoust. Soc. Am. **118**(3), 1677–1688 (2005)

Torreira, F., Adda-Decker, M., Ernestus, M.: The Nijmegen corpus of casual French. Speech Commun. **52**, 201–212 (2010)

Tuller, B., Kelso, J.A.S.: The production and perception of syllable structure. J. Speech Hear. Res. **34**, 501–508 (1991)

Turco, G., Fougeron, C., Audibert, A.: Que nous apprennent les gros corpus sur l'harmonie vocalique en français? In: Proceedings of the 2016 JEP-TALN-RECITAL, Paris, pp. 571–579 (2016)

Walker, D.C.: French Sound Structure. University of Calgary Press, Calgary (2001)

Whalen, D.H.: Coarticulation is largely planned 7/3. J. Phon. **18**, 3–35 (1990)

Word-Initial Irregular Phonation
as a Function of Speech Rate and Vowel
Quality in Hungarian

Alexandra Markó[1,2(✉)], Andrea Deme[1,2], Márton Bartók[1,2],
Tekla Etelka Gráczi[1,3], and Tamás Gábor Csapó[1,4]

[1] MTA–ELTE "Lendület" Lingual Articulation Research Group,
Budapest, Hungary
[2] Department of Phonetics, Eötvös Loránd University, Budapest, Hungary
[3] Research Institute for Linguistics, Hungarian Academy of Sciences,
Budapest, Hungary
[4] Department of Telecommunication and Media Informatics,
Budapest University of Technology and Economics, Budapest, Hungary
{marko.alexandra,deme.andrea}@btk.elte.hu,
bartokmarton@gmail.com,
graczi.tekla.etelka@nytud.mta.hu, csapot@tmit.bme.hu

Abstract. We examined vowel-initial irregular phonation in real words as a function of vowel quality, backness and height, and speech rate in Hungarian. We analyzed two types of irregular phonation: glottalization and glottal stop. We found that open vowels elicited more irregular phonation than mid and close ones, but we found no effect of the backness. The frequency of irregular phonation was lower in fast than in slow speech. Inconsistently with the claims of earlier studies, the relative frequency of glottalization to glottal stops was not influenced by speech rate in general. However, while /i/ was produced with a relatively higher ratio of glottal stops in fast speech, the open vowels showed the widely documented tendency of being realized with relatively less glottal stops under the same conditions.

Keywords: Irregular phonation · Glottal stop · Glottalization · Vowel height · Vowel backness · Speech rate

1 Introduction

Irregular phonation is an umbrella term in the literature, which covers several realization types of irregularity in vocal fold vibration. Other terms like *laryngealization, glottalization, creaky voice,* etc. are also used, and in several cases they refer to only more or less similar domains of irregularity. Based on their formal characteristics, some authors use more accurate definitions for the subtypes, (e.g., Batliner et al. 1993, Dilley et al. 1996), while in several studies the concept of irregularity is introduced in a more intuitive manner. Considering the terminological variability in the literature, it is crucial that we clarify our use of terms in the present work. We refer to irregularity in general using the term *irregular phonation,* and we adopt the definition formed by Surana and

© Springer Nature Switzerland AG 2018
Q. Fang et al. (Eds.): ISSP 2017, LNAI 10733, pp. 134–145, 2018.
https://doi.org/10.1007/978-3-030-00126-1_13

Slifka (2006: p. 693): "A region of phonation is an example of irregular phonation if the speech waveform displays either an unusual difference in time or amplitude over adjacent pitch periods that exceeds the small-scale jitter and shimmer differences, or an unusually wide-spacing of the glottal pulses compared to their spacing in the local environment, indicating an anomaly with respect to the usual, quasiperiodic behavior of the vocal folds."

In the present study, we investigate two easily distinguishable types of irregular phonation. For one of these phenomena we apply the term *glottalization* (covering several possible subtypes) to refer to cases where irregularity can be observed as consecutive periods in voicing differing evidently in terms of duration, amplitude, or both. The second phenomenon is the unique glottal gesture which we refer to as *glottal stop* (Fig. 1.). As Esling and Harris (2005) pointed out, a single isolated burst (or set of aperiodic bursts) "may differ in timing from creaky voice but not in laryngeal configuration" (372). According to their interpretation, glottalization and glottal stop are articulated similarly; and several studies do also consider these phenomena various forms of the same glottal behavior (see e.g., Kohler 2001, Malisz et al. 2013).

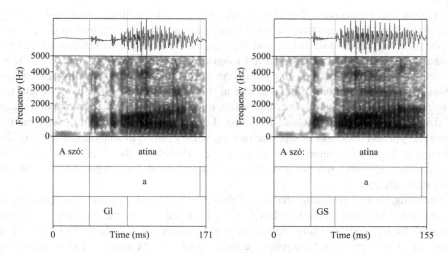

Fig. 1. Examples of glottalization (left) and glottal stop (right)

Irregular phonation serves prosodic functions in typologically unrelated languages, e.g., American English (Dilley et al. 1996), Czech, Spanish (Bissiri et al. 2011), German, Polish (Kohler 2001; Malisz et al. 2013), Hungarian (Markó 2013) and others. The occurrence of irregularity may be influenced by several factors (see some of these below), and it shows high inter- and intraspeaker variability (e.g., Dilley et al. 1996; Redi and Shattuck-Hufnagel 2001).

Kohler (2001: pp. 282–285) defined four types of irregular phonation (which he labelled as *glottalization* covering "the glottal stop and any deviation from canonical modal voice") as follows. (1) Vowel-related glottalization phenomena which signal the boundaries of words or morphemes. (2) Plosive-related phenomena which occur as reinforcement or even replacement of plosives. (3) Syllable-related phenomena which characterize syllable types along a scale from a glottal stop to glottalization (e.g., Danish stød). (4) Utterance-related phenomena which comprise (i) phrase-final relaxation of phonation, and (ii) truncation glottalization, i.e., utterance-internal tensing of phonation at utterance breaks.

Vowel-initial irregular phonation (a specific case of type (1) above) was analyzed in several studies. Malisz et al. (2013) examined the conditioning effect of speech style (speech vs. dialogue), presence of prominence, phrasal position (initial vs. medial), speech rate, word type, preceding segment, and following vowel height on the occurrence frequency of word-initial glottalization in Polish and German. They concluded -among others- that vowels bearing prominence were more frequently marked glottally (in both languages), faster rates reduced glottal marking in general, especially the occurrence frequency of glottal stops, but faster rates increased the relative frequency of occurrence of glottalization. They also found that low vowels were more frequently glottalized in both languages than non-low vowels; however it must also be noted that speech rate and vowel quality factors were not systematically varied in this study.

Lancia and Grawunder (2014) used pseudo-words to facilitate vowel-initial irregular phonation, and to analyze the conditioning factors of vowel height (high vs. low: /i/ vs. /a/), the presence of stress, and the place of articulation of the preceding consonants. They concluded, that retracted tongue body (i.e., low-back tongue position) favors the production of irregular phonation (particularly strongly in unstressed syllables). It should be noted, however, that in the cited study the high-front /i/ and the low-back /a/ were compared, thus the results might have revealed an interaction effect of the features vowel height and backness. Lancia and Grawunder did not treat the vowel height and backness as separate factors, as they aimed to analyze the effect of the maximally retracted tongue position using /a/. Therefore, the cited results are not informative regarding the possible independent effect of the tongue height and the front-back dimension.

In Hungarian a systematic analysis of the effect of speech rate and vowel quality in vowel-initial irregular phonation has not been carried out so far; in addition, to the authors' knowledge, a study considering the interaction of these factors is also nonexistent for any other languages either. Moreover, in earlier studies regarding vowel-related irregularity in Hungarian (e.g., Markó 2013), glottalization and glottal stops were not treated separately. Therefore no data is available on the relative frequency of these two types either.

In the present study we investigated the effect of vowel height and backness, and speech rate, using a balanced speech material, which was also well controlled in terms of speech rate. In the analysis, two types of irregular phonation (glottalization and glottal stop) were also taken into account.

Based on previous results for other languages we addressed the following questions. Is irregular phonation more frequent in word-initial vowels in Hungarian if (i) the speech rate is slow (as opposed to fast); (ii) the vowel is back (as opposed to front); (iii) the vowel is open (as opposed to close or close-mid)? (iv) Do glottal stops occur less frequently in fast speech, while the relative amount of glottalization increases? We hypothesized that the occurrence frequency of irregular phonation in general is higher in slow speech than in fast speech. Furthermore, we assumed that back vowels elicit irregular phonation in a higher ratio than front ones both in slow and fast speech, and that open vowels favor irregular phonation more than non-open (close or close-mid) ones. Finally, we also assumed that faster speech rate reduces the amount of glottal stops, but increases the relative frequency of glottalization.

2   Material and Method

2.1   Material and Experiment Design

The test material consisted of disyllabic Hungarian pronominal adverbs: *innen* /inːɛn/ 'from here'; *onnan* /onːɒn/ 'from there'; *ennek* /ɛnːɛk/ 'for this'; *annak* /ɒnːɒk/ 'for that'. These adverbs start with four different vowel qualities which vary both in the vowel height (close /i/ vs. mid /o/ vs. open /ɛ ɒ/) and the backness (back /o ɒ/ vs. front /i ɛ/; while backness also co-varies with lip spreading) (see Fig. 2). (It is important to note here that in the present study by the introduction of /ɛ/ as an open vowel we used a "simplified" feature set along the vowel height dimension to which the system described by Szende (Fig. 2) and the similar "acoustic openness" of /ɛ/ and /ɒ/ provided the basis.)

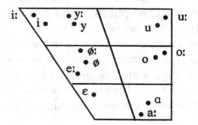

Fig. 2. The Hungarian vowel inventory (Szende 1994: p. 92) (Please note, that even though in this system Szende labels the /ɒ/ quality without the lip-rounding gesture, i.e., as /ɑ/, the rounded /ɒ/ is a more generally accepted transcription of the sound in question).

These target words were embedded in the following phrase: *Mondd:* [target word] *kell.* 'Say: [target word] needed'. All of the word-initial target vowels bear sentencial accent on the first syllable (given that word stress is fixed on the first syllable in Hungarian).

The stimuli were presented on a computer screen. Each trial consisted of two display screens: first the introductory part (*Mondd:*) was showed to the participant, then the target item (target word + *kell*) was displayed. The participants' task was to read aloud the target item, but not the introductory part.

In order to elicit speech rate differences between the conditions, the timing of the display screens was manipulated. In the "slow" speech condition, each display screen appeared for 1500 ms resulting in 3000 ms for one trial in total (including the introductory part and the "target word + *kell*" construction). In the "fast" speech condition the timer was set to 500 ms, resulting in 1000 ms for one trial in total. (During the recordings several other timer settings were also applied, and the setting to serve as the "fast" condition was selected posterior to the recordings on the basis of the speakers' ability to produce the items properly, i.e., separately and without errors.) As the timing of the introductory part reflected also the timing of the target item (target word + *kell*), it enabled the speakers to prepare for the production of the latter one.

The trials were ordered into blocks: within each block all the four different target words occurred in a randomized order once, and these blocks were repeated 5 times consecutively for each ("slow" and "fast") condition. First the "slow" condition, then the "fast" condition was recorded in the case of every participant. 4 vowel qualities/ target words × 5 repetitions × 2 speech rate conditions, i.e., 40 vowels per speaker were recorded. The recordings were made in a sound-treated booth, using a tie-clip omnidirectional condenser microphone and an external soundcard.

2.2   Participants

Previous results revealed that Hungarian female speakers tend to produce irregular phonation more frequently than male speakers (see e.g., Markó 2013); therefore in the present study only female speakers were included; all 33 of them were university students, and native speakers of Hungarian, who reported no hearing or speech deficits. In order to ensure that the "slow" and the "fast" conditions differentiate properly (i.e., they may be differentiated by a conceptually sound value), a threshold for the speech rate difference was introduced (see below, in 2.4).

This threshold was not exceeded by the data in the case of 15 participants, thus finally in the main analysis 18 speakers' material was involved. The speakers' age ranged between 19 and 34 years, with a mean of 24.9 years.

In the "slow" condition 359 vowels were analyzed (as one speaker mispronounced one target word), while in the "fast" condition the number of analyzed vowels was 360.

2.3   Annotation

The "target word + *kell*" construction, the word-initial vowel, and the irregular phonation at the beginning of the word-initial vowel were labelled manually in Praat (Boersma and Weenink 2016). The vowel qualities were labelled automatically (on the basis of the stimuli order), and then the quality label assignment was also confirmed manually as correct by the annotators (two of the authors of the present paper) auditorily. In the case of mispronunciation or any other errors involving the production of the vowel of interest, the vowel was excluded from the material. Vowel boundaries were defined on the basis of the F_2 trajectory.

The labeling of the irregular phonation was performed in accordance with the methodology proposed by previous studies (e.g., Dilley et al. 1996; Bőhm and Ujváry 2008) in which visual (waveform and spectrogram) and auditive information was combined. A given vowel was labelled as irregular if (i) its first consecutive f_0-periods differed evidently in terms of duration, amplitude, or both (these cases were marked as glottalization, see Fig. 1, left), or if (ii) one (or more) glottal stop was observed at the beginning of the vowel (these cases were marked as glottal stop, see Fig. 1, right) (see Dilley et al. 1996). Those cases where glottal stops were followed by creaky voice were labelled as glottalization. The occurrences of irregular phonation were noted by one listener, who marked the entire corpus in Praat. In the case of ambiguous occurrences the pitch curve and pulses shown by Praat (standard settings) were also taken into consideration.

We analyzed the ratio of vowels produced with irregular phonation with respect to vowel quality, vowel height and backness, and speech rate. We also determined and compared the ratio of glottalized vowel occurrences to the number of vowels realized with glottal stops in the two speech rate conditions.

2.4 Control of Speech Rate

Regarding that one of the aims of the present study was making a comparison between "slow" and "fast" speech, it was inevitable to carefully control the difference of the speech rate between these two conditions. To achieve this goal, first we manipulated the timing of the display screens (see above). And second, we also set a perceptually motivated threshold for the speech rate differences.

In the case of Hungarian, the just noticeable difference (JND) for speech tempo has not been studied so far; however, there are JND data for other languages which may be taken as a reasonable reference for Hungarian as well. For instance, for Dutch speech fragments Quené (2007) found 5% JND for artificially increased/decreased speech rate differences, while he also noted that this value may be an overestimation, and that in the case of everyday communicative situations, the JND is probably lower.

As in our case the number of the phonemes per item was constant, to calculate speech rate differences, not the speech sound per duration values, but only the item durations were measured in the two speech rate conditions.

The target item durations were measured and compared both in "slow" and "fast" conditions speakerwise. However, as in our study five repetitions of each item were analyzed and averaged, we decided to apply a higher threshold of 10% to account for the expected reduction in duration variability. As mentioned above, there were 18 speakers who differentiated their speech rates by more than 10% on average; therefore these 18 speakers' data were included in the main analysis. The duration difference between the two conditions ranged between 9.6% and 28.5% speakerwise, and the mean of differences was $16.8 \pm 5.5\%$. The overall item duration was 852 ± 107 ms in the "slow" condition, and 705 ± 69 ms in the "fast" condition (for all 18 speakers).

2.5 Statistical Analyses

Three 2-way repeated measures ANOVAs were performed with the factors *vowel quality* and *speech rate*, *vowel backness* and *speech rate*, and *vowel height* and *speech rate*, and a confidence level set to 95%.

3 Results

3.1 Vowel-Initial Irregular Phonation as a Function of Vowel Quality and Speech Rate

The ratio of vowels produced with irregular phonation (pooled over speakers) as a function of vowel quality and speech rate is presented in Fig. 3.

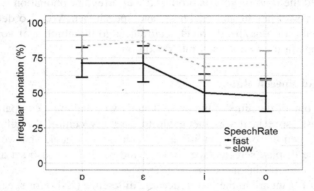

Fig. 3. The ratio of vowels produced with any kind of irregular phonation as a function of vowel quality and speech rate (mean + 95% CI)

In the "slow" condition the ratio of vowel-initial irregular phonation in the front and close /i/ was 68.6 ± 20.7%, while in the "fast" condition it was 50.0 ± 30.1% of all cases. In the case of the front and open vowel /ɛ/ these ratios were 86.7 ± 18.1% in the "slow" and 71.1 ± 29.3% in the "fast" conditions. The back and mid vowel /o/ was produced with irregular phonation 70.0 ± 24.0% of all cases in the "slow" and 47.8 ± 25.8% of all cases in the "fast" condition. Finally, the back and open /ɒ/ showed 83.3 ± 18.5% irregular occurrences in the "slow" and 71.1 ± 24.9% in the "fast" condition.

The ANOVA showed significant main effects of both *speech rate* ($F(1, 17) = 17.38$, $p < 0.001$) and *vowel quality* ($F(3, 51) = 10.56$, $p < 0.001$), but the interaction of these factors turned out to be non-significant.

3.2 Vowel-Initial Irregular Phonation as a Function of Vowel Backness and Speech Rate

The ratio of vowels produced with irregular phonation as a function of vowel backness and speech rate is shown in Fig. 4. The back /o/ and /ɒ/ and the front /ɛ/ and /i/ vowels were produced with irregular phonation in a similar ratio both in the "slow" and the "fast" conditions. In the "slow" condition front vowels showed 77.6 ± 21.2% ratio of irregular occurrences, while back vowels showed 76.7 ± 22.1% of all cases. In the "fast" condition the ratios were 60.6 ± 31.2% and 59.4 ± 27.7%, respectively. Regarding this comparison, statistical analysis showed a significant difference between the *speech rate* conditions ($F(1, 68) = 14.58$, $p < 0.001$) but not between the front and back vowel groups.

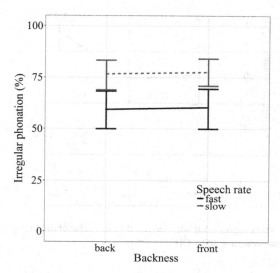

Fig. 4. The ratio of vowels produced with any kind of irregular phonation as a function of vowel backness and speech rate (mean + 95% CI)

3.3   Vowel-Initial Irregular Phonation as a Function of Vowel Height and Speech Rate

The ratio of vowel realizations with irregular phonation in terms of vowel height and speech rate are shown in Fig. 5. The close vowel /i/ was produced with irregular phonation 68.6 ± 20.7% of all tokens in the "slow" and 50.0 ± 30.1% of all tokens in the "fast" condition. The mid vowel /o/ showed vowel-initial irregular phonation 70.0 ± 24.0% of all cases in the "slow" and 47.8 ± 25.8% in the "fast" condition. The open vowels /ɒ/ and /ɛ/ were produced with irregular phonation at the highest ratio: 85.0 ± 18.1% in the "slow" and 71.1 ± 26.8% in the "fast" condition.

According to the ANOVA, there was no significant interaction between *speech rate* and *vowel height*, but both factors had a significant main effect: vowel height: $F(2, 34) = 13.20$, $p < 0.001$; speech rate: $F(2, 17) = 17.28$, $p < 0.001$.

3.4   Ratio of Glottalization and Glottal Stops as a Function of Vowel Quality and Speech Rate

The ratio of occurrences of glottal stops and glottalization as a function of vowel quality and speech rate are presented in Fig. 6. Relative to the number of all vowel realizations (not displayed on Fig. 6.) both in the "slow" and the "fast" conditions the percentage of glottalization (42.6% and 35.0% of all occurrences, respectively) exceeded the percentage of glottal stops (34.5% and 25.0% of all occurrences,

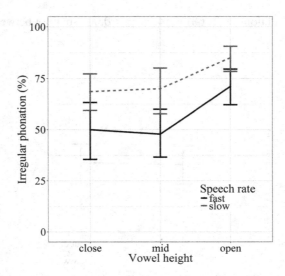

Fig. 5. The ratio of vowels produced with any kind of irregular phonation as a function of vowel height and speech rate (mean + 95% CI)

respectively). Relative to the number of all irregular occurrences both in the "slow" and the "fast" conditions the percentage of glottalization (54.9% and 56.0% of all irregular occurrences, respectively) exceeded the percentage of glottal stops (45.1% and 43.4% of all irregular occurrences, respectively) as well (see Fig. 6).

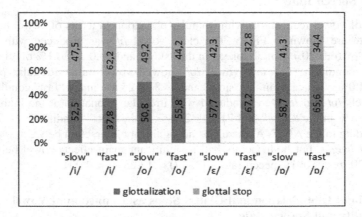

Fig. 6. The ratio of the two types of irregular phonation (relative to all irregular occurrences) as a function of vowel quality and speech rate

The ratio of glottalization and glottal stops were close to equal in the case of /i/ in the "slow" condition, while in the "fast" condition the ratio of glottal stops was well above the ratio of glottalization. For the "slow" condition the pattern was very similar in the case of /o/; in the "fast" condition, however, glottalization was relatively more frequent than glottal stops. In the case of /ɛ/ and /ɒ/, the ratio of glottalization exceeded the ratio of glottal stops in both of the conditions. Although we observed differences in the glottalization to glottal stop ratio between the two conditions in three of the four analyzed vowels, the close /i/ showed the opposite tendency with the change in speech rate, than that observed in the case of open vowels. The direction of the change in the mid /o/ was the same as in the case of the open vowels, but the degree of change was smaller.

4 Discussion and Conclusion

Our first hypothesis, claiming that the occurrence frequency of irregular phonation is higher in slow than in fast speech, has been confirmed. Our second and third hypotheses were partially confirmed by the data: even though vowel backness did not have an effect on vowel-initial irregular phonation, we showed that open vowels favor irregular phonation more than mid and close ones both in slow and fast speech. Our fourth assumption claiming that faster speech rates reduce the relative amount of glottal stops, while increasing the frequency occurrence of glottalization was not verified, since glottalization was more frequent in both speech rates in general. However, to some extent, at the different vowel heights studied, different tendencies were found.

Considering that in the present study the effect of phonetic position, vowel quality, and speech rate were strictly controlled and investigated in real words, we can conclude that open vowels tend to elicit more irregular phonation than mid and close ones, irrespective of the vowel backness. We can also conclude that the frequency of irregular phonation tends to be lower in fast than in slow speech (or at least in speech accelerated under laboratory conditions). Relative frequency of glottalization to glottal stops in phrase-initial vowel-initial position did not appear to be influenced by speech rate in general, which itself was inconsistent with the claims of earlier studies (e.g., Malisz et al. 2013). However, taking the analyzed vowels separately into account, we observed that the behavior of the close /i/ was opposite to that of the open /ɒ/ and /ɛ/. While the open vowels showed the widely documented tendency of being realized with relatively less glottal stops in fast speech, /i/ was produced with a relatively higher ratio of glottal stops under the same conditions. This result suggests that the vowel height has an effect not only on the frequency of irregular phonation, but also on the manner of its realization in the case of word- and phrase-initial vowels.

As already mentioned in the Introduction section, Lancia and Grawunder (2014) reported that the effect of laryngealization was found to be weaker when the tongue was more fronted (in the case of /i/) and that laryngealization was produced with a retraction of the tongue (in the case of /a/). They explained this finding as a result of an inter-action between the tongue position and the laryngeal settings, namely that higher tongue positions also increase the vertical position of the larynx, which then modifies the mode of phonation (lower positions leading to more lax, while higher positions leading to more tense phonation). However, as we also pointed out, the vowel height and backness features were not considered as separate factors in the cited study, thus their possible individual effects were not revealed. In the present study, we aimed to tease apart these factors, and we showed that it is the vowel height feature that exhibits an effect on the occurrence of irregular phonation, as the low back /ɒ/ and the low front /ɛ/ vowels showed similarly high ratio of irregular occurrences, irrespective of their backness feature. Interpreted in the light of the argumentation of Lancia and Grawunder (2014), we thus conclude that it may be the vowel height that is the most influential vocalic feature on the vertical position of the larynx, and thus indirectly it is also the most influential with respect to the mode of phonation (at least as far as the occurrence of word-initial irregular phonation is considered).

This claim is also supported by previous articulatory studies that suggest, that front close unrounded vowels (e.g., /i/, /e/) have a higher larynx position than back close, close-mid or open rounded vowels (as /u/, /o/, /ɒ/) (Hess 1998, Hoole and Kroos 1998, Demolin et al. 2002). However, as Esling (2005: p. 23) points out, from an articulatory perspective, the difference between "openness" categories should also be considered fundamentally different in "front" than in "back" vowels, as in "front" vowels it is indeed the difference in jaw opening, while in "back" vowels it is raising in the case of "close", and retraction of the tongue in the case of "open" vowels. And on this basis, Esling (2005) also suggests that due to the connections of the tongue muscles to other speech organs, the "back open" /ɒ/-like vowels, should or might exhibit the strongest connection to (and the greatest effect on) the laryngeal settings via the direct link of the hyoid bone. Even though we also found that high vowels are less likely to facilitate the occurrence of irregular phonation, the latter prediction, i.e. the special effect of low/back vowels, was not confirmed by our data. To clarify the question, if the backness feature of the vowels does indeed not have a relevant effect on the occurrence frequency of irregular phonation, we plan to extend our study by the analysis of articulatory data of the vowels at hand.

Acknowledgement. The authors are grateful to Gergely Varjasi for his valuable help in recruitment of the participants and in conducting the experiments.

References

Batliner, A., Burger, S., Johne, B., Kiessling, A.: MÜSLI: a classification scheme for laryngealizations. In: Working Papers, Prosody Workshop, pp. 176–179. Schweden, Lund (1993)

Bissiri, M.P., Lecumberri, M.L., Cooke, M., Volín, J.: The role of word-initial glottal stops in recognizing English words. In: Proceedings of Interspeech 2011, pp. 165–168, Florence (2011)

Boersma, P., Weenink, D.: Praat: doing phonetics by computer. Version 6.0.17 (2016). http://www.praat.org/

Bőhm, T., Ujváry, I.: Az irreguláris fonáció mint egyéni hangjellemző a magyar beszédben. Beszédkutatás **2008**, 108–120 (2008)

Demolin, D., Hassid, S., Metens, T., Soquet, A.: Real-time MRI and articulatory coordination in speech. Comptes Rendus – Biologies **325**(4), 547–556 (2002)

Dilley, L., Shattuck-Hufnagel, S., Ostendorf, M.: Glottalization of word-initial vowels as a function of prosodic structure. J. Phon. **24**, 423–444 (1996)

Esling, J.H.: There are no back vowels: The laryngeal articulator model. Can. J. Linguist. **50**, 13–44 (2005)

Esling, J.H., Harris, J.: States of the glottis: An articulatory phonetic model based on laryngoscopic observations. In: Hardcastle, W.J., Beck, J. (eds.) A Figure of Speech: A Festschrift for John Laver, pp. 347–383. Lawrence Erlbaum Associates, Mahwah (2005)

Hess, S.A.: Pharyngeal Articulations. Ph.D. dissertation, UCLA (1998)

Hoole, P., Kroos, C.: Control of larynx height in vowel production. In: Proceedings of the 5th Conference on Language Processing (ICSLP), vol. 2, pp. 531–534 (1998)

Kohler, K.J.: Plosive-related glottalization phenomena in read and spontaneous speech. A stød in German? In: Grønnum, N., Rischel, J. (eds.) To Honour Eli Fischer-Jørgensen, pp. 174–211. Kopenhagen, Reitzel (2001)

Lancia, L., Grawunder, S.: Tongue-larynx interactions in the production of word initial laryngealization over different prosodic contexts: a repeated speech experiment. In: Fuchs, S., Grice, M., Hermes, A., Lancia, L., Mücke, D. (eds.) Proceedings of the 10th (ISSP), pp. 245–248, Cologne (2014)

Malisz, Z., Żygis, M., Pompino-Marschall, B.: Rhythmic structure effects on glottalisation: a study of different speech styles in Polish and German. Lab. Phon. **4**(1), 119–158 (2013)

Markó, A.: Az irreguláris zönge funkciói a magyar beszédben. ELTE Eötvös Kiadó, Budapest (2013)

Quené, H.: On the just noticeable difference for tempo in speech. J. Phon. **35**, 353–362 (2007)

Redi, L., Shattuck-Hufnagel, S.: Variation in the realization of glottalization in normal speakers. J. Phon. **29**, 407–429 (2001)

Surana, K., Slifka, J.: Acoustic cues for the classification of regular and irregular phonation. In: INTERSPEECH-2006, paper 1755-Mon3FoP.1 (2006)

Szende, T.: Illustrations of the IPA: Hungarian. JIPA **24**(2), 91–94 (1994)

Effects of Entering Tone on Vowel Duration and Formants in Nanjing Dialect

Yongkai Yang and Ying Chen[(✉)]

School of Foreign Studies, Nanjing University of Science and Technology,
Nanjing, China
yongkai_yang@foxmail.com, ychen@njust.edu.cn

Abstract. There are five tones in Nanjing Dialect, including four open-syllable tones and one entering tone with a syllable-final glottal stop, which is not found in Beijing Mandarin. Each checked syllable normally corresponds to an open syllable with the same vowel but different tones. This study focuses on the acoustic comparisons of the two kinds of syllables to discuss the acoustic discrepancies between checked and open syllables in Nanjing Dialect, specifically with regards to vowel quality. The acoustic parameters include the duration, the first formant (F1), the second formant (F2), and the third formant (F3) of the vowels. The results of vowel duration indicate that the entering tone is still the shortest among the five tones. The relatively higher F1, F2 and F3 in the entering tone suggest the effect of a glottal stop coda on the vowel quality.

Keywords: Nanjing Dialect · Entering tone · Open syllables · Vowel duration
Formant structure

1 Introduction

Chao [1] identified five tones in Nanjing Dialect: T1—*yinping* (mid-falling), T2—*yangping* (mid-rising), T3—*shangsheng* (low-level), T4—*qusheng* (high-level), and T5—*rusheng* (high-short). The relative pitch values of the five tones according to the five-scale annotation [2] are T1 (31), T2 (13), T3 (22), T4 (44), and T5 (5). *Rusheng* in Chinese literally means "entering tone", referring to checked-syllable tones; *yinping*, *yangping*, *shangsheng* and *qusheng* are four open-syllable tones. A syllable with the entering tone in Nanjing Dialect ends with a glottal stop [ʔ], which makes the laryngeal and mouth muscle tense and obstructs airflow in the vocal tract. Sun [3] concludes that the characteristics of checked syllables in Nanjing Dialect are high pitch, short duration, glottal stop coda, and muscular tension.

Previous studies on the basis of phonologists' perceptual experience imply that, due to the influence of modern Standard Mandarin, the distinctive features of checked syllables in Nanjing Dialect are diminishing. The signs of this change include the loss of the glottal stop and the prolonged duration of the checked syllable [4, 5]. Based on the predictions and conclusions of the previous phonological studies, the current study investigates the acoustics of the entering tone vs. its corresponding open-syllables in Nanjing Dialect, specifically to explore the effect of this potential sound change on vowel production.

© Springer Nature Switzerland AG 2018
Q. Fang et al. (Eds.): ISSP 2017, LNAI 10733, pp. 146–157, 2018.
https://doi.org/10.1007/978-3-030-00126-1_14

The experiment was designed with tone-vowel sequences involving the five tones and different vowels for the acoustic measurements of duration, F1, F2, and F3 of the vowels. We are particularly interested in (1) whether the synchronic data indicate shorter duration for the entering tone compared to other tones, and (2) whether the formant structure varies with and without the glottal stop coda. Considering the phonetic and phonological differences between younger and older speakers' production and bimodal distribution of male and female production found in Chen and Wiltshire [6], the present study also takes account of the effects of age and gender.

2   Methods

2.1   Participants

Eighteen native speakers of Nanjing Dialect participated in the current study. All participants reported no speech and hearing problems. There were three age groups: younger (aged from 21 to 25), mid-age (aged from 41 to 49) and older (aged from 62 to 72). Each group had three male and three female speakers. All participants were born and raised in Nanjing and use Nanjing Dialect in their daily communication, though they were taught to speak standard Mandarin at school. To avoid accent diversity within the dialect, all participants were recruited from the urban area of Nanjing.

2.2   Stimuli

Participants were required to read the words in Nanjing Dialect in Table 1. The target words for recording were chosen from *Nanjing Dialect Dictionary* [7]. Six vowels /a, ə, o, i, u, y/ were included for the minimal pairs of open vs. checked syllables. Each vowel was preceded with the same initial consonant and overridden with five tones, including four open syllables and one checked syllable with the final glottal stop. A total of 30 high-frequency words, shown in Table 1, were selected and embedded in a carrier phrase of "×, 这个字是 × 字" (×, *zhe ge zi shi × zi*. '×, this character is × character') for recording.

2.3   Recording

Participants were recorded in the sound-attenuated booth at the School of Foreign Studies, Nanjing University of Science and Technology. Each participant was recorded with 90 words (6 vowels × 5 tones × 3 times) embedded in the carrier sentence in a random order with normal loudness and speech rate. A Marantz PMD661 professional recorder and a Shure SM10A-CN head-worn microphone were used to record the stimuli in a mono channel with 44,100 Hz sampling rate. The sounds were digitized on an SD card to save on the computer.

2.4   Analysis

Acoustic data were collected by the phonetic analysis software Praat, first using ProsodyPro (Version 6.1.1), a Praat script for prosody analysis to track the F0 trajectory

Table 1. Target words in Nanjing Dialect

T1	T2	T3	T4	T5
叉 [tʂʰa³¹] 'fork'	查 [tʂʰa¹³] 'check'	衩 [tʂʰa²²] 'pants'	岔 [tʂʰa⁴⁴] 'turnoff'	插 [tʂʰaʔ⁵] 'insert'
奢 [ʂə³¹] 'luxury'	蛇 [ʂə¹³] 'snake'	舍 [ʂə²²] 'abandon'	舍 [ʂə⁴⁴] 'dorm'	舌 [ʂəʔ⁵] 'tongue'
拖 [tʰo³¹] 'drag'	驼 [tʰo¹³] 'carry'	妥 [tʰo²²] 'proper'	唾 [tʰo⁴⁴] 'saliva'	脱 [tʰoʔ⁵] 'take off'
妻 [tɕʰi³¹] 'wife'	奇 [tɕʰi¹³] 'strange'	起 [tɕʰi²²] 'up'	气 [tɕʰi⁴⁴] 'air'	七 [tɕʰiʔ⁵] 'seven'
夫 [fu³¹] 'husband'	扶 [fu¹³] 'support'	府 [fu²²] 'mansion'	富 [fu⁴⁴] 'rich'	服[fuʔ⁵] 'cloth'
区 [tɕʰy³¹] 'district'	渠 [tɕʰy¹³] 'canal'	取 [tɕʰy²²] 'fetch'	趣 [tɕʰy⁴⁴] 'fun'	屈 [tɕʰyʔ⁵] 'bend'

for tone [8], and then using FormantPro (Version 1.4), another Praat script for vowel analysis [9]. Only the first "×" in the carrier sentence was extracted as the target sound for the acoustic analysis in this study. The default setting of Maximum Formant (MF) in the Praat script was 5000 Hz for male and 5500 Hz for female. However, after manual observation of the formant structure of all participants' production, some female speakers' MF setting was adjusted to 6000 Hz and 6500 Hz, while some male speakers' MF setting to 4500 Hz in order to track accurate values of vowel formants. Descriptive data of duration, F1, F2, and F3 of the vowels did not show large variations (standard deviation) in the averages across speakers in each subject group. Therefore, the mean duration, F1, F2, and F3 of the vowels were directly analyzed in four sets of repeated measures ANOVA. In each set of ANOVA, there were two within-subject factors, tone (five levels: T1 = (31), T2 = (13), T3 = (22), T4 = (44), T5 = (5)) and vowel (six levels: /a/, /e/, /o/, /i/, /u/, /y/), and two between-subject factors, age (three levels: younger, mid-age and older) and gender (two levels: male and female).

3   Results

3.1   F0 Contours

Before doing statistical analysis for duration and formants, the F0 contour of each tone type was first examined. F0 values at 10 even-interval points were extracted from each syllable and then averaged for tone, age and gender across speakers. The vowel duration was also averaged for tone, age and gender. The 10 averaged F0 values were plotted over the actual duration in Fig. 1.

Figure 1 shows wider pitch range of female speakers than male speakers and slightly longer duration of older speakers than mid-age and younger speakers. Figure 1 clearly indicates that T5 is the highest and shortest one among the five tones in Nanjing Dialect. The raw data also indicate 252 out of 324 (78%) T5 tokens demonstrated at

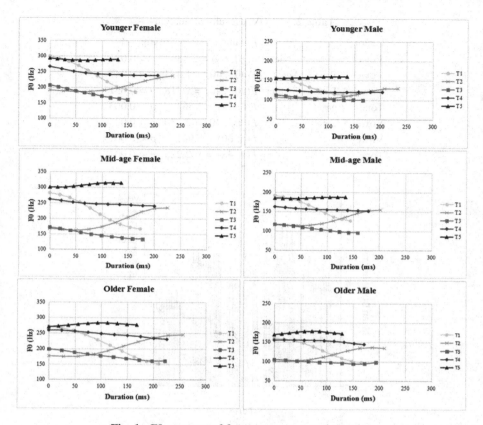

Fig. 1. F0 contours of five tones over actual duration

least one cycle of creaky voice in the waveforms due to laryngealization of the glottal stop coda. However, it did not sharply pull down the F0 contour.

3.2   Duration

The ANOVA results of vowel duration with significant interactions and main effects are indicated in Table 2. Post hoc paired-sample t-tests for duration comparisons between the entering tone and the open-syllable tones indicate that T5 is always significantly shorter $(p < 0.001)$[1] than the other four tones under the condition of vowel type. Figure 2 illustrates these differences.

---

[1] Here the reference of p value for multiple comparisons was not adjusted since Perneger [10] argued that this kind of downward adjustment is overly conservative.

Table 2. ANOVA results of duation

Factors	F value	Sig.
Vowel * Age * Gender	$F(10, 60) = 2.134$	$p = 0.022$
Vowel * Tone * Age	$F(40, 240) = 1.453$	$p = 0.047$
Vowel * Tone * Gender	$F(20, 240) = 1.718$	$p = 0.031$
Tone * Age	$F(8, 48) = 3.563$	$p = 0.003$
Vowel * Tone	$F(20, 240) = 2.130$	$p = 0.004$
Vowel	$F(5, 60) = 8.941$	$p < 0.001$
Tone	$F(4, 48) = 102.225$	$p < 0.001$

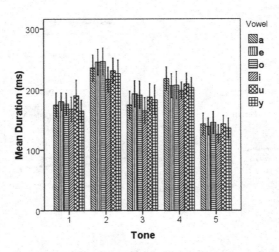

Fig. 2. Duration differences by vowel and tone

3.3   F1

Significant interactions and main effects in the ANOVA results of F1 are indicated in Table 3.

Table 3. ANOVA results of F1

Factors	F value	Sig.
Vowel * Age * Gender	$F(10, 60) = 2.134$	$p = 0.022$
Vowel * Tone * Age	$F(40, 240) = 1.453$	$p = 0.047$
Vowel * Tone * Gender	$F(20, 240) = 1.718$	$p = 0.031$
Tone * Age	$F(8, 48) = 3.563$	$p = 0.003$
Vowel * Tone	$F(20, 240) = 2.130$	$p = 0.004$
Vowel	$F(5, 60) = 8.941$	$p < 0.001$
Tone	$F(4, 48) = 102.225$	$p < 0.001$

Figure 3 demonstrates F1 differences by tone and vowel and is separated into two panels by gender since gender has a main effect while age does not.

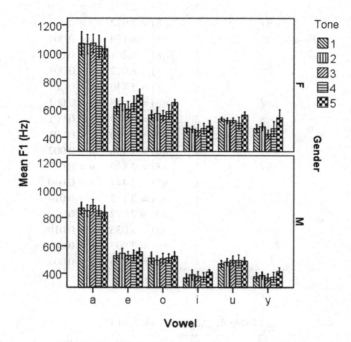

Fig. 3. F1 differences by vowel, tone and gender

Table 4 shows the post hoc paired-sample *t*-test results of F1 comparisons between the entering tone and other tones.

Within-subjects analysis indicates a higher F1 for the entering tone than for other tones, except for the vowel /a/. For /a/, the F1 of the entering tone is lower than other tones, but only significant for T1 and T3 in the production of male speakers.

3.4   F2

Table 5 shows the ANOVA results of F2 with significant interactions and main effects.

Table 4. F1 comparisons in *t*-test

Gender	Vowel	Tone	Tone	*T* value	Sig.
Male	/a/	5	1	$t(8) = -3.414$	$p = 0.009$
			3	$t(8) = -4.258$	$p = 0.003$
	/e/	5	3	$t(8) = 2.987$	$p = 0.017$
	/o/	5	2	$t(8) = 3.274$	$p = 0.011$
	/i/	5	1	$t(8) = 2.935$	$p = 0.019$

(*continued*)

Table 4. (*continued*)

Gender	Vowel	Tone	Tone	T value	Sig.
			4	$t(8) = 2.894$	$p = 0.020$
	/y/	5	3	$t(8) = 2.632$	$p = 0.030$
Female	/e/	5	1	$t(8) = 6.159$	$p < 0.001$
			2	$t(8) = 8.201$	$p < 0.001$
			3	$t(8) = 8.012$	$p < 0.001$
			4	$t(8) = 5.830$	$p < 0.001$
	/o/	5	1	$t(8) = 8.648$	$p < 0.001$
			2	$t(8) = 6.089$	$p < 0.001$
			3	$t(8) = 7.180$	$p < 0.001$
			4	$t(8) = 4.466$	$p = 0.002$
	/u/	5	2	$t(8) = 2.830$	$p = 0.022$
			3	$t(8) = 2.444$	$p = 0.040$
			4	$t(8) = 3.135$	$p = 0.014$
	/y/	5	1	$t(8) = 2.872$	$p = 0.021$
			2	$t(8) = 3.033$	$p = 0.016$
			3	$t(8) = 3.907$	$p = 0.004$
			4	$t(8) = 5.206$	$p = 0.001$

Table 5. ANOVA results of F2

Factors	F value	Sig.
Vowel * Tone * Age * Gender	$F(40, 240) = 1.565$	$p = 0.022$
Tone * Age * Gender	$F(8, 48) = 2.772$	$p = 0.015$
Vowel * Tone * Age	$F(40, 240) = 1.471$	$p = 0.042$
Vowel * Tone * Gender	$F(20, 240) = 1.849$	$p = 0.017$
Vowel * Age	$F(10, 60) = 3.334$	$p = 0.002$
Vowel * Gender	$F(5, 60) = 13.845$	$p < 0.001$
Tone * Age	$F(8, 48) = 2.664$	$p = 0.017$
Tone * Gender	$F(4, 48) = 4.074$	$p = 0.006$
Vowel * Tone	$F(20, 240) = 3.900$	$p < 0.001$
Vowel	$F(5, 60) = 1013.16$	$p < 0.001$
Tone	$F(4, 48) = 13.544$	$p < 0.001$
Age	$F(2, 12) = 4.597$	$p = 0.033$
Gender	$F(1, 12) = 4.597$	$p < 0.001$

Based on the main effects, F2 differences by tone, vowel, gender and age are illustrated in Fig. 4.

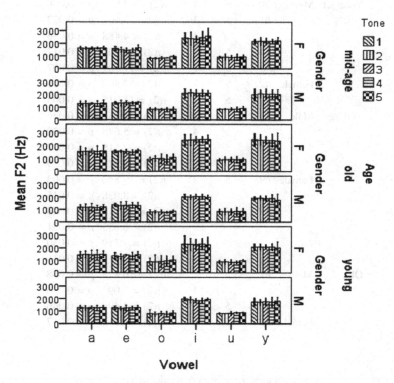

Fig. 4. F2 differences by vowel, tone, age and gender

Table 6 shows the post hoc t-test results of F2 comparison between the entering tone and open-syllable tones.

Table 6 and Fig. 4 indicate that the F2 of the entering-tone vowels is generally higher than that of other tones. T-test results reveal that the effect is strong on /e/, /u/, and /o/ in the production of older generations and on /i/, /y/, /a/ and /o/ in the production of the younger generation.

3.5   F3

The ANOVA results of F3 with significant interaction and main effects are shown in Table 7.

Table 6. F2 comparisons in t-test

Age	Gender	Vowel	Tone	Tone	T value	Sig.
Younger	Male	/o/	5	4	$t(2) = 12.850$	$p = 0.006$
		/i/	5	3	$t(2) = 6.667$	$p = 0.022$
				4	$t(2) = 4.488$	$p = 0.046$
		/u/	5	2	$t(2) = 7.601$	$p = 0.017$
		/y/	5	1	$t(2) = 11.507$	$p = 0.007$
	Female	/a/	5	1	$t(2) = 7.809$	$p = 0.016$
		/o/	5	1	$t(2) = 14.015$	$p = 0.005$
Mid-age	Male	/e/	5	1	$t(2) = 5.159$	$p = 0.036$
				4	$t(2) = 5.645$	$p = 0.030$
		/u/	5	3	$t(2) = 7.515$	$p = 0.017$
				4	$t(2) = 4.928$	$p = 0.039$
	Female	/e/	5	2	$t(2) = 19.848$	$p = 0.003$
				3	$t(2) = 8.038$	$p = 0.015$
		/o/	5	1	$t(2) = 18.110$	$p = 0.003$
				2	$t(2) = 6.855$	$p = 0.021$
				3	$t(2) = 4.319$	$p = 0.050$
				4	$t(2) = 7.917$	$p = 0.016$
Older	Male	/o/	5	2	$t(2) = 10.873$	$p = 0.008$
				4	$t(2) = 5.095$	$p = 0.036$
		/u/	5	3	$t(2) = 4.496$	$p = 0.046$
	Female	/e/	5	1	$t(2) = 7.569$	$p = 0.017$

Table 7. ANOVA results of F3

Factors	F value	Sig.
Vowel * Tone * Gender	$F(20, 240) = 1.789$	$p = 0.022$
Vowel * Gender	$F(5, 60) = 8.162$	$p < 0.001$
Tone * Age	$F(8, 48) = 2.328$	$p = 0.034$
Tone * Gender	$F(4, 48) = 4.837$	$p = 0.002$
Vowel * Tone	$F(20, 240) = 6.226$	$p < 0.001$
Vowel	$F(5, 60) = 461.946$	$p < 0.001$
Tone	$F(4, 48) = 13.918$	$p < 0.001$
Gender	$F(1, 12) = 62.529$	$p < 0.001$

Figure 5 shows the F3 differences by tone, vowel, and gender. Age was not included since there was no main effect of age.

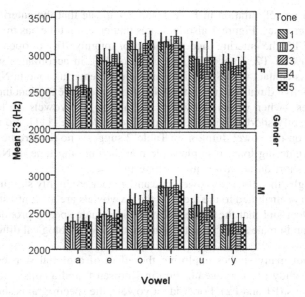

Fig. 5. F3 differences by vowel, tone, and gender

Table 8 shows the post hoc *t*-test results of F3 comparisons between the entering tone and open-syllable tones.

Table 8. F3 comparisons in *t*-test

Gender	Vowel	Tone	Tone	T value	Sig.
Male	/u/	5	1	$t(8) = 2.958$	$p = 0.018$
Female	/o/	5	2	$t(8) = 3.850$	$p = 0.005$
			3	$t(8) = 5.559$	$p = 0.001$
	/y/	5	3	$t(8) = 2.338$	$p = 0.048$
			4	$t(8) = 2.363$	$p = 0.046$

Table 8 indicates that, when comparing the entering tone with other tones, only rounded vowels /y/, /u/, and /o/ show significant differences in F3.

4 Discussion

This study focuses mainly on how the glottal stop at the syllable coda position affects the duration and formants of vowels with the entering tone (i.e., T5). Since syllable structures of minimal pairs with a same vowel but different open-syllable tones are always the same, we are not particularly interested in comparing the duration and formants of the vowels among open-syllable tones (i.e., T1-4). Instead, our discussion will specifically compare the effect of the entering tone on vowel duration and formants with that of open syllables.

The data on vowel duration in Figs. 1 and 2 indicate that the entering tone is still shorter than other tones. Figure 1 also indicates that entering tone has the highest pitch. The duration of T5, the entering tone, accounts for roughly 70% of open-syllable tones (79% of T1, 60% of T2, 76% of T3, and 67% of T4), in accordance with Song [4]. Although there is contact with Standard Mandarin, the entering tone in Nanjing Dialect retains its distinctive duration, suggesting Nanjing Dialect is experiencing an evolution of entering tones—when words end with a glottal stop, the vowels are lengthened but are still significantly shorter than those in open-syllable tones [11]. No main effect of age and gender on the vowel duration in Table 2 suggests no diachronic change in the duration of the entering tone. It is plausible that the entering tone in Nanjing Dialect could retain its short duration for quite a long time.

Unsurprisingly, the effects of vowel type and gender are highly significant on all the formants, which is attributed to the fact that the six vowels are all contrastive phonemes in Nanjing Dialect and that the vowel formants in female speech production are generally higher than in male speech production due to the physiological difference of their vocal tracts.

There are not many studies exploring the effect of a glottal stop coda on vowel quality, but the study of pharyngealization in Khoisan found a considerable raising of the lower formants (F1 and F2). For strident vowels, the spectrograms show even more upward displacement of F1 and F2, but the lowering of F3 is observed in pharyngealized vowels [12, 13].

The current study has found a higher F1 with the entering tone than with other tones, except for tones occurring with /a/. As high F1 indicates low tongue position, it is conceivable that the lowering of the glottis is also concomitant with the lowering of the jaw and the tongue. The reason why F1 of /a/ in the male speakers' production shows a different pattern from other vowels in Table 4 may be that the tongue position of /a/ is already low and a sudden laryngeal tenseness leads to the uplift of tongue root. This impact does not apply to /e/, /o/, /i/, /u/, and /y/ because the tongue positions of these vowels are relatively higher and there is some space for tongue lowering.

Unlike F1, F2 value is influenced by all four factors—vowel, tone, group, and gender. Figure 4 and Table 6 suggest that the laryngeal tenseness of the entering tone affects the backness of vowel, as a higher F2 indicates a more front tongue position. This tonal effect is stronger for back vowels /o/ and /u/ than for front vowels /i/ and /y/ and stronger for /e/, which is intrinsically not as front as /i/ and /y/. One possibility is that laryngeal tenseness leads to a more front tongue position, but the effect is not that strong for vowels that are already in very front positions due to limitation of the oral space.

A lower F3 indicates a rounder lip shape in the articulation. All significant differences of F3 were found in round vowels /o/, /u/ and /y/. For /i/, /e/ and /a/, there were no significant differences between the entering tone and other tones. These results indicate that the post-vowel glottal constriction only affects rounded vowels and not unrounded vowels. F3 was higher in /o/, /u/, and /y/ with the entering tone than with other tones as in Fig. 5 and Table 8.

According to the perturbation theory, constriction of the vocal tract around the larynx and pharynx raises F1 and F2. However, constriction of the vocal tract at the larynx raises

F3 but at the pharynx, lowers F3 [14]. This clearly explains why F1 and F2 were raised but F3 was lowered in the study of pharyngealization in Khoisan [12, 13], whereas in the current study, all three formants were raised due to laryngealization.

5 Conclusion

The current study reveals that the glottal stop coda not only influences vowel duration, but also vowel quality. The discrepancy in the formant structure of vowels between the entering tone and other tones does not suggest the dropping of the glottal stop coda in Nanjing Dialect.

Acknowledgement. This work was supported by the National Science Foundation of China (61573187) and Humanity and Social Science Foundation of Ministry of Education of China (18YJA740007).

References

1. Chao, Y.: The phonology of Nanjing Dialect. J. Sci. **13**, 1005–1036 (1929)
2. Chao, Y.: A system of tone letters. Le Maître Phonétique **45**, 24–27 (1930)
3. Sun, H.: A discussion on tones of Nanjing Dialect. J. Nanjing Xiaozhuang Univ. **19**, 34–40 (2003)
4. Song, Y.: On experiment of entering tones in Nanjing Dialect. J. Sch. Chin. Lang. Cult. Nanjing Norm. Univ. **11**, 166–172 (2009)
5. Sun, H.: A discussion on light tone and entering tone in Nanjing Dialect. J. Jiangsu Second. Norm. Univ. **17**, 67–68 (2001)
6. Chen, S., Wiltshire, C.: Tone realization in younger versus older speakers of Nanjing Dialect. In: Jing-Schmidt, Z. (ed.) Increased Empiricism: Recent Advances in Chinese Linguistics, pp. 147–170. Benjamins, Amsterdam and Philadelphia (2003)
7. Liu, D.: Nanjing Dialect Dictionary. Jiangsu Education Publisher, Nanjing (1995)
8. Xu, Y.: ProsodyPro — a tool for large-scale systematic prosody analysis. In: TRASP 2013, Aix-en-Provence, France, pp. 7–10 (2013)
9. Xu, Y.: FormantPro.praat (2007–2015). http://www.phon.ucl.ac.uk/home/yi/FormantPro/
10. Perneger, T.: What's wrong with Bonferroni adjustments. BMJ **316**, 1236–1238 (1998)
11. Yan, T.: The physiological characteristics of syllables with entering tones in Chinese dialects — the diachronic change of entering tones. Lang. Linguist. China **1**, 523–537 (1992)
12. Catford, J.: Pharyngeal and laryngeal sounds in Caucasian language. In: Bless, D.M., Abbs, J.H. (eds.) Vocal Folds Physiology: Contemporary Research and Clinical Issues, pp. 344–350. College Hill Press, San Diego (1983)
13. Lagefoged, P., Maddieson, I.: The Sounds of the World's Languages, pp. 310–313. Blackwell Publishers, Oxford (1996)
14. Johnson, K.: Acoustic and Auditory Phonetics, pp. 137–141. Blackwell Publishers, Oxford (2012)

Which Factors Can Explain Individual Outcome Differences When Learning a New Articulatory-to-Acoustic Mapping?

Eugen Klein[1(✉)], Jana Brunner[1], and Phil Hoole[2]

[1] Humboldt-Universität zu Berlin, Berlin, Germany
{eugen.klein,jana.brunner}@hu-berlin.de
[2] Ludwig-Maximilians-Universität München, Munich, Germany
hoole@phonetik.uni-muenchen.de

Abstract. Speech motor learning is characterized by inter-speaker outcome differences where some speakers fail to compensate for articulatory and/or auditory perturbations. Hypotheses put forward to explain these differences entertain the idea that speakers employ auditory and sensorimotor feedback differently depending on their predispositions or different acuity traits. A related idea implies that individual speakers' traits may further interact with the amount of auditory and somatosensory feedback involved in the production of a specific speech sound, e.g. with the degree of the tongue-palate contact. To investigate these hypotheses, we performed two experiments with an identical group of Russian native speakers where we perturbed vowel and fricative spectra employing identical experimental designs. In both experiments we observe compensatory efforts for all participants. However, among our participants we find neither compelling evidence for individual feedback preferences nor for consistent speaker-internal patterns of the learning outcomes in the context of vowels and fricatives. We suggest that a more plausible explanation for our results is provided by the idea that fricatives and vowels exhibit different degrees of complexity of the articulatory-to-acoustic mapping.

Keywords: Real-time spectral fricative perturbation · Formant perturbation
Articulatory-to-acoustic mapping · Sensorimotor learning

1 Introduction

Previous articulatory and auditory perturbation studies demonstrate that speakers can quickly relearn articulatory configurations they use to produce a speech sound in an unperturbed condition. For instance, Fowler and Turvey [1] examined participants' productions of vowels when a bite block was inserted between their teeth. The authors found that speakers were able to adapt to these perturbations instantly without practice and produce acoustic outputs equivalent to their unperturbed speech. However, in a study by Savariaux et al. [2] when speakers' lips were blocked with a tube during the production of the French [u] only six out of 11 speakers were able to partially compensate for the labial perturbation and only one speaker compensated completely by changing the constriction location from a velo-palatal to a velo-pharyngeal region.

© Springer Nature Switzerland AG 2018
Q. Fang et al. (Eds.): ISSP 2017, LNAI 10733, pp. 158–172, 2018.
https://doi.org/10.1007/978-3-030-00126-1_15

Similar compensatory variability is also observed across and within other studies [e.g., 3–5]. The outcome of such studies comes down to the questions of whether speakers can compensate for the perturbation, and to what extent they can 'restore' the intended speech sound to its form before the perturbation was applied. A range of hypotheses has been put forward to explain the individual outcome differences which are observed in speech motor learning tasks.

Lametti et al. [6], for instance, promote the hypothesis that speakers have individual preferences for articulatory or auditory feedback signals to control their speech production. To empirically support their claim, the authors investigated participants in different experimental conditions. In one condition, the authors perturbed speakers' jaw trajectories by means of a robotic arm without altering the corresponding speech acoustics, in another they shifted only speakers' auditory feedback without applying articulatory perturbation. In the third condition, the authors applied both types of perturbation simultaneously. The authors found a negative correlation between the amount of articulatory and auditory compensation, which means that speakers who compensated for articulatory perturbations at the same time compensated to a lesser degree for auditory shifts.

However, the hypothesis that speakers exhibit individual preferences for articulatory or auditory feedback conflicts with previous observations made by Ghosh et al. [7] who investigated somatosensory and auditory acuity, the degree to which speakers are sensitive to changes in articulatory and auditory signals. These authors found that both types of acuity positively correlated with each other as well as with the magnitude of produced sibilant contrasts. Furthermore, auditory acuity has been shown to positively correlate with adaptation magnitude in auditory perturbation of vowels [8] as well as in articulatory perturbation of sibilants [9].

To investigate Lametti et al.'s hypothesis further, we conducted two auditory perturbation experiments to compare speakers' reaction to real-time shifts of the Russian close-central unrounded vowel /ɨ/ and the palatalized fricative /sʲ/. This vowel vs. fricative comparison was motivated by the assumption that speakers rely on different amounts of auditory and somatosensory feedback during the production of vowels compared to consonants [10], since the latter involve stronger physical contact between speakers' tongue and the palate. To reduce the potential confounding effect of individual feedback preferences claimed by Lametti et al., we recruited the same participants for both experiments.

In both experiments, the spectral properties of the target segments were perturbed in opposing directions depending on the stimulus in which the targets were embedded. In the case of the vowel /ɨ/, the second formant (F2) was perturbed downwards or upwards depending on the preceding consonant (/d/ or /g/). In the case of the fricative /sʲ/, the whole spectrum was perturbed downwards or upwards causing the center of gravity (COG) to increase or decrease. We chose this particular perturbation design for both experiments as studies have previously shown that speakers are able to simultaneously use multiple articulatory configurations to produce equivalent acoustic outputs [5]. Thus, the bidirectional perturbation ensures that acoustic changes observed in speakers' production are indeed ascribable to compensation rather than being caused by other unrelated factors, such as fatigue. In our study, we expand the bidirectional paradigm to fricatives and compare individual learning outcomes in tasks involving auditory perturbation of vowels and fricatives.

Assuming that individual speakers prefer different feedback channels to monitor their own speech production, we should observe a comparable amount of compensation for each participant within and across both experiments (vowels and fricatives) irrespective of the applied perturbation direction. That is, if a speaker is able to compensate for the perturbation in one direction during the first experiment, s/he should be able to perform in a similar manner for the opposite direction during the first experiment, as well as for both perturbation directions during the second experiment. However, if we find asymmetric compensation patterns and/or systematic differences in compensatory behavior between vowels and fricatives this would suggest that additional factors may influence speakers' compensatory behavior.

2 Method

2.1 Participants

18 native monolingual speakers of Russian (14 females) without reported speech, language, or hearing disorders participated in the study. The participant group consisted of exchange students and young professionals living in Berlin. The mean age of the group was 24.6 years (range 20–32) and the participants had spent on average 2.8 years (range 1–4) in Germany at the time of the recordings. The study was approved by the local ethics committee and all participants gave their written consent to participate in both experiments.

2.2 Procedure

Each experimental session consisted of two experiments (vowel and fricative perturbation) and was recorded in a sound attenuated booth. Participants were seated in front of a computer monitor which served to display stimuli and additional task descriptions.

Each of the experiments consisted of 210 trials and lasted for about 25 min. Participants completed both experiments with a short break in between. During the baseline phase, where no perturbation was applied, participants had to produce the four CV syllables [di], [dɨ], [gɨ], and [gu] during the first experiment, and the six CVC words [les], [ves], [lesʲ], [vesʲ], [veʃʲ], and [leʃʲ] during the second experiment (cf. Table 1). After the baseline phase, three shift phases followed during which the spectral properties of the syllables [dɨ] and [gɨ] in the first experiment and the words [lesʲ] and [vesʲ] in the second experiment were perturbed. The additional stimuli recorded during the baseline phase served the purpose to identify any specifics of speakers' sound maps that could interact with their compensation magnitude.

For the production of CV syllables, participants were asked to prolong the vowel to maximize the time they were exposed to the perturbation. For the second experiment, participants were asked to produce each CVC syllable with a neutral intonation, in a normal speech tempo to improve online tracking of the fricative. To keep the speech

Table 1. (a) The experimental stimuli for the Experiment 1 and (b) the Experiment 2. The shaded cells contain the perturbed stimuli.

A	[gu]	[di]	[gi]	[gu]

B	[les]	'wood'	[lesʲ]	'climb'	[leʃʲ]	'bream'
	[ves]	'weight'	[vesʲ]	'whole'	[veʃʲ]	'thing'

amplitude equal across both experiments, participants were provided with a real-time graphic display of the microphone gain and asked to keep a certain intensity range.

Participants' speech was recorded with a Beyerdynamic Opus 54 neck-worn microphone, perturbed in real-time, and fed back via EARTONE 3A insert earphones.

2.3 Acoustic Manipulations

For real-time perturbation of the spectral frequencies we employed AUDAPTER, a C++ real-time signal processing application executable within MATLAB [11]. The original and perturbed audio signals were digitized and saved with a sampling rate of 16 kHz for vowels and of 32 kHz for fricatives, respectively. Beside audio recordings, AUDAPTER stored data files containing the formant vectors (F1 and F2) tracked during each vowel production.

During the first experiment, AUDAPTER tracked the formants during the vowel production and shifted F2 in real-time. F2 was decreased in one experimental /Ci/syllable and increased in the other during the three shift phases (Fig. 1). The combination of the perturbation direction with the initial consonant of the CV syllable was counterbalanced between the 18 participants to control for any potential effect of the place of articulation. The amount of F2 perturbation increased from 220 to 520 Hz in 150 Hz steps over the three shift phases. The perturbation amount did not change within each shift phase.

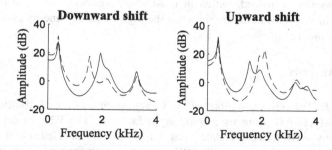

Fig. 1. Example LPC-spectra of the original (solid lines) and perturbed (dashed lines) vowel segments during the last shift phase of the Experiment 1.

During the second experiment, we employed AUDAPTER's pitch shifting facilities. When applied to voiceless fricative spectra, this manipulation results in an overall spectral shift such that the COG of the fricative is increased or decreased depending on the direction of the pitch shift. During the shift phases of the second experiment, pitch was decreased by five semitones during the fricative in the word [vesʲ] and increased by five semitones during the fricative in the word [lesʲ] (Fig. 2). The pitch shift was applied exclusively to the fricative portion of each stimulus and did not affect the first two segments. The perturbation amount remained the same throughout the three shift phases.

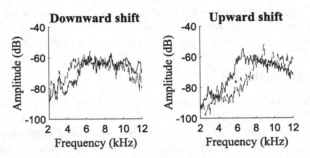

Fig. 2. Example power spectra of the original (solid lines) and perturbed (dashed lines) fricative segments during the shift phases of the Experiment 2.

2.4   Real-Time Segment Tracking

To identify the onset and the offset of the fricative, AUDAPTER performed a real-time analysis of the speech signal's short-time root-mean-square (RMS) and RMS ratio curves (Fig. 3a). While RMS is an acoustic intensity measure, RMS ratio is an indicator of high frequency intensity present in the signal and is computed by dividing smoothed RMS curve by a high-pass filtered RMS curve. Operating with a set of heuristic rules, AUDAPTER enabled pitch shifting when a predefined high frequency energy threshold was crossed, which in most cases coincided with the onset of the fricative, and turned it off immediately when the high frequency energy curve fell again under the threshold (Fig. 3b).

2.5   Data Analysis

Experiment 1. The onset and offset of the vowel produced on each trial were labeled manually in MATLAB using its graphic input facilities. The F2 formant vector was extracted from AUDAPTER's data files based on the labeled onset and offset boundaries. Subsequently, the middle 50% portion of each formant vector was used to compute the F2 means for each experimental trial. To pool F2 data across all participants, raw frequencies produced during each shift phase were normalized by subtracting each participant's mean F2 frequency produced during the baseline phase for the respective syllable.

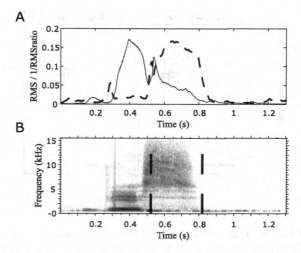

Fig. 3. Example of an experimental trial of the second experiment: (a) RMS (solid line) and RMS ratio curve (dashed line) of the speech signal. (b) Fricative onset and offset (dashed lines) tracked by AUDAPTER overlaid over a spectrogram of the speech signal.

Experiment 2. After the onset and offset of the fricatives produced on each trial were labeled manually, the COG vector was extracted from the fricative portion of each response. For that, the speech signal was high pass filtered at 1 kHz with a 6th order Butterworth filter and power spectral densities (PSD) were computed over the middle 50% portion of the fricative. The COG was computed for the accumulated and time averaged power spectra. To pool COG data across all participants, raw COG frequencies were normalized by subtracting each participant's mean COG frequency produced during the baseline phase for the respective syllable.

3   Results

For both experiments, we will start the review of the results by summarizing the initial frequencies produced by the participants during the baseline phase. After that, we will turn to the presentation of the compensatory behavior observed during the shift phases. Here, we will discuss changes in F2 and COG frequencies produced by participants across the three shift phases. Finally, we will compare compensatory behavior observed among participants across both experiments and discuss different compensatory patterns.

3.1   Experiment 1

Initial Sound Map. The mean formant frequencies produced by all participants during the baseline phase for the syllables [di], [dɨ], [gɨ], and [gu] are summarized in Fig. 4.

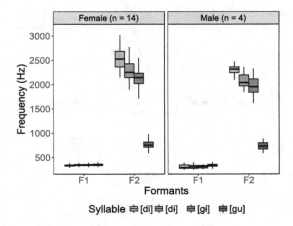

Fig. 4. Average F1 and F2 frequencies produced during the baseline phase (no perturbation) for the four syllables [di], [dɨ], [gɨ], and [gu]. The data is split by participants' gender.

Overall, the observed formants were consistent with previous descriptive studies of Russian vowel space [12].

A linear mixed-effects model [13] was fitted for each of the two formants including the produced syllable and the interaction between the syllable and gender as fixed effects and the frequency as dependent variable. Random intercepts were modeled for every participant. P-values were obtained with the lmerTest package [14].

The model suggested that the average difference in F1 between [dɨ] and [di] was 9.4 Hz ($t = 5.55, p < 0.05$) and 2.92 Hz between [gɨ] and [dɨ] ($t = 1.73, p > 0.05$). The F1 difference between [gu] and [gɨ] was 4.65 Hz ($t = 2.76, p < 0.05$).

The model for F2 indicated a significant within-speaker difference between F2 values for [dɨ] and [di] (-253.43 Hz, $t = -19.78, p < 0.05$). Also, the average F2 value for [gu] was significantly different from [gɨ] (-1356.94 Hz, $t = -106.16, p < 0.05$). The average F2 value was lower for [gɨ] compared to [dɨ] (-149.54 Hz, $t = -11.69$, $p < 0.05$).

All formant frequencies produced by male participants were lower compared to formants produced by female participants. These differences were all significant except for F1 and F2 differences in [gu].

Compensatory Behavior. Here, we examine compensatory effects observed in the vowel [ɨ] during the three shift phases. Based on the general observation that speakers produce formant values opposing the perturbation, we expected the F2 values produced during the baseline phase for the two target syllables to drift apart over the course of the three shift phases. On the other hand, we did not expect to observe any compensatory changes in the F1 frequency.

To quantify changes in F1 and F2, we fitted a generalized additive mixed model (GAMM) [15] which included random smooths for each participant by trial number and a factor smooth for trial number with intercept difference for the perturbation direction. This allowed us to capture intra- and inter-individual variability of the compensatory behavior without making any restrictive assumptions regarding its temporal or spatial characteristics.

As expected, there were no meaningful changes of the F1 frequency during the shift phases. The average compensatory effects in F2 are summarized in Fig. 5a. The model suggested that the average F2 frequency was significantly different between trials with downward and upward perturbation (-94.45 Hz, $t = -38.98$, $p < 0.05$). In Fig. 5b, we further see that this difference increased to approximately 300 Hz by the end of the experiment. In line with observations from previous perturbation studies, the modeling of the data suggested that participants strongly differed in their individual compensation magnitudes (Fig. 5c). The overall directional trend of the random smooths further suggests that the compensation magnitude was on average higher for the upward perturbation.

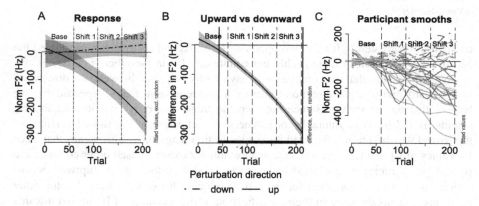

Fig. 5. (a) Average compensatory effects (excluding random participant effects) in F2 for downward and upward perturbation during the three shift phases. (b) The average difference in F2 between the opposing compensatory effects. Black thick lines denote the region of significance. (c) Random model smooths for each participant.

Based on individual smooths, we identified three subgroups among our participants exhibiting different learning patterns. Eight participants ("symmetrical learners") compensated for F2 perturbation in both syllables containing the vowel [ɨ] in the opposite direction of the applied perturbation. As expected, the average compensation had a magnitude of approximately 30–45% independent of the perturbation magnitude (Fig. 6a). However, it appears that the compensation was overall slightly stronger during the upward perturbation. Another five participants ("asymmetrical learners") compensated for F2 perturbation essentially only when it was shifted upwards, but not when it was shifted downwards. Consequently, the F2 values for both syllables containing /ɨ/ diverged over the course of the shift phases for the asymmetrical learners, too. However, when F2 in /ɨ/ was shifted downwards, asymmetrical learners produced these vowels with approximately same absolute F2 values across all learning phases (Fig. 6b). When F2 in /ɨ/ was shifted upwards, asymmetrical learners compensated up to 60% of the applied perturbation. Five participants ("non-learners") failed to

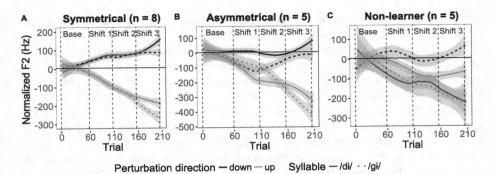

Fig. 6. Normalized F2 values produced across all phases of the Experiment 1. Participants' data is divided based on their learning pattern: (a) symmetrical learners, (b) asymmetrical learners, (c) non-learner.

compensate consistently for the two opposing F2 shifts and seem to have followed the perturbation direction in one syllable but counteracted it in the other (Fig. 6c).

We hypothesize that the compensatory asymmetry observed between the downward and upward perturbation directions mirrors a phonetic asymmetry. Specifically, we believe that the phonetic distance between the target vowel [ɨ] and its neighboring sounds [i] and [u] is accountable for listeners' classification of the shifted versions of that vowel which in turn influenced the compensation magnitude. This means that since F2 values of [i] and [ɨ] were on average substantially closer to each other compared to [ɨ] and [u], participants classified instances of [ɨ] shifted towards [i] (upward perturbation) as errors and corrected for them by producing lower F2 values. On the other hand, participants deviated in their classification of the instances of [ɨ] shifted towards [u] (downward perturbation) and either compensated for the shifts (symmetrical learners) or tended to ignore them (asymmetrical learners).

3.2   Experiment 2

Initial Sound Map. To examine the average COG values produced by all participants during the baseline phase, we first collapsed the data of the six CVC words into three sound categories according to their final segment. The results of this aggregation are summarized in Fig. 7. Overall, the COG frequencies of the fricatives [s], [sʲ], [ʃ] were consistent with previous descriptive studies of Russian [12].

We fitted a linear-mixed model with COG as dependent variable and fixed terms for the sound category and the interaction between sound category and gender as well as random intercept for each participant. The model suggested that COG was significantly different between [s] and [sʲ] (-837.74 Hz, $t = -10.42$, $p < 0.05$) and between [sʲ] and [ʃ] (-3076.51 Hz, $t = -38.28$, $p < 0.05$). Male speakers produced on average lower COG values for all three categories, however, the interaction between the sound category and gender was not significant for any of the fricative sounds.

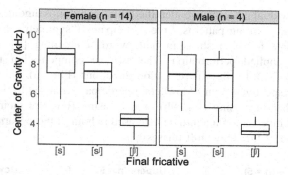

Fig. 7. Averaged COG produced during the baseline phase (no perturbation) for the three fricatives [s], [sʲ], [ʃ]. The data is split by participants' gender.

Compensatory Behavior. To quantify the COG changes that appeared during perturbation of the fricative [sʲ], we fitted a GAMM which included random smooths for each participant by trial number and a factor smooth for trial number with intercept difference for the perturbation direction. The average compensatory effects are summarized in Fig. 8a. As we can see, the average COG frequency produced for the two words containing the fricative [sʲ] diverged in the course of the three shift phases and reached approximately 110 Hz by the end of the experiment (Fig. 8b). The model suggested that the average difference between downward and upward perturbation was significant (-43.64 Hz, $t = -2.395$, $p < 0.05$), even though, the effect size was much smaller compared to the first experiment investigating F2 perturbation. Furthermore, the compensation variability among participants was higher in contrast to the first experiment as revealed by individual smooths fitted by the model (Fig. 8c). There was no overall directional trend of the individual participant smooths which would suggest a stronger compensatory reaction to one of the two perturbation directions.

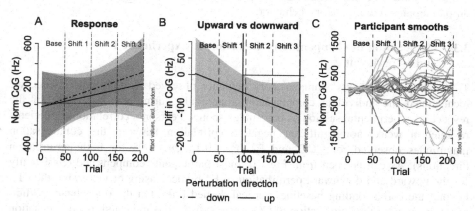

Fig. 8. (a) Average compensatory effects (excluding random participant effects) in COG for downward and upward perturbation during the three shift phases. (b) The average difference in COG between the opposing compensatory effects. Black thick lines denote the region of significance. (c) Random model smooths for each participant.

Based on individual smooths, we identified three subgroups among our participants exhibiting different learning patterns. Five participants ("symmetrical learners") were able to compensate for pitch-shifts in both words containing [sʲ] in the opposite direction of the applied perturbation. The average compensation magnitude was approximately 10–20% for both perturbation directions (Fig. 9a). On the other hand, nine ("uppers") respectively four ("downers") participants reacted to both perturbation directions by either increasing (Fig. 9b) or decreasing (Fig. 9c) their mean spectral frequency. The compensation magnitude increased in both of these learner groups up to 80% in the course of the three shift phases.

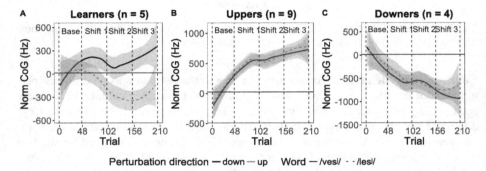

Fig. 9. Baseline-normalized COG values produced across all phases of the Experiment 2. Participants' data is divided based on their learning pattern: (a) symmetrical learners, (b) uppers, (c) downers.

In contrast to the F2 perturbation, where the phonetic distance between the shifted sound and its neighbors influenced the compensation magnitude, the neighboring sounds [s] and [ʃʲ] of the shifted fricative [sʲ] do not seem to have a systematic influence on participants' compensatory behavior.

3.3   Comparison of Compensatory Behavior in Experiment 1 and Experiment 2

To correlate participants' compensatory behavior observed in both experiments, we recalculated the individual compensation magnitude during the last phase of the respective experimental session as percentage scores. Pearson's correlation coefficients revealed a weak, non-significant negative correlation between the compensation magnitudes observed for COG and F2 ($r = -0.15$, $p > 0.05$ for both perturbation directions). As can be seen from Fig. 10, most participants compensated consistently for the upward and downward perturbation of F2 by decreasing or increasing their F2 beyond the corresponding baseline (vertical dashed line). On the other hand, participants' reaction to the perturbation of COG was indistinct as indicated by compensation scores which fluctuated around the corresponding baseline (horizontal dashed line) independent of the perturbation direction.

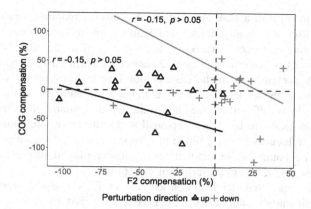

Fig. 10. Correlation of compensatory magnitudes observed in Experiment 1 (vowels) and Experiment 2 (fricatives) measured as percentual change in COG and F2, respectively. The data is spilt by the perturbation direction.

To comprehend the regularities as well as the differences in individual compensatory strategies, we summarized the results of both experiments in a contingency table (Table 2). There were three participants who were able to compensate consistently for both perturbation directions both in vowels and in fricatives (top left cell). However, for the rest of the participants the compensatory strategy they employed in the first experiment did not predict their compensatory behavior in the second experiment. Moreover, we were not able to identify any generalizable tendencies among the different learning patterns.

Table 2. Compensatory strategies for all participants across both experiments.

F2	COG			
	Learners	Uppers	Downers	Total
Symmetrical	3	2	3	**8**
Asymmetrical	1	3	1	**5**
Non-learners	1	4	–	**5**
Total	**5**	**9**	**4**	**18**

4 Discussion and Conclusion

In this paper, we presented results from two bidirectional auditory perturbation studies of the Russian close-central unrounded vowel [ɨ] and the palatalized fricative [sʲ] which were conducted with the same participants.

The results of the first experiment examining compensation for F2 perturbation are consistent with a previous study by Rochet-Capellan and Ostry [5] who show that speakers are able to quickly relearn their initial articulatory-to-acoustics mapping in the F1 domain. However, five out of 18 speakers who participated in the first experiment were not able to acquire a new articulatory-to-acoustics mapping reacting inconsistently to F2 shifts. Specifically, these participants seem to have compensated for F2 perturbation in one perturbation direction but followed it in the other. If we consider Lametti et al.'s [6] hypothesis that compensation for perturbation is driven by individual feedback preferences, in the case of non-learners' results we would have to assume that during the experiment they referenced the auditory and somatosensory feedback channels to different degrees depending on the syllable they had to produce. This explanation seems to be rather implausible. Furthermore, previous findings that auditory acuity influences the magnitude of compensation for formant perturbation [8] similarly cannot account for the fact that participants employed different compensation strategies for the two different perturbation directions.

Instead, we argue that the five non-learners in Experiment 1 were not able to identify the articulatory parameters they needed to adjust in order to achieve the intended acoustic goal. In other words, these speakers were not able to translate consistently the perceived acoustic errors into appropriate corrective articulatory movements. The idea that the non-learners indeed perceived the acoustic errors caused by the perturbation is evident from the fact that their production of the target vowel [ɨ] changed substantially during the experiment. This line of explanation is consistent with the results of the second experiment investigating bidirectional spectral shifts of the fricative [sʲ].

In the second experiment, we again observed a behavior split among the participants concerning their ability to compensate for the applied perturbation. However, compared to the first experiment, where five speakers were not able to consistently react to the bidirectional shifts, the number of participants who failed to do so in the second experiment was much higher with 13 participants (uppers and downers). Analogous to the five non-learners in the first experiment, uppers and downers seem to have compensated for the perturbation in one direction but followed it in the other. This means that these participants must have perceived the acoustic errors caused by the perturbation as they reacted to them by substantially increasing or decreasing the COG in their fricative production. As in the case of the non-learners in the first experiment, the failure of these 13 participants to compensate differently depending on the perturbation direction remains unexplainable either by participants' auditory and sensorimotor feedback preferences or by their different acuity traits. Again, we are left with the explanation that the 13 participants who failed to achieve the intended acoustic goal during fricative perturbation were not able to identify the relevant articulatory parameters they needed to adjust to do so.

Comparing the compensation strategies that each participant applied during the first and the second experiment, we could not identify any speaker-internal correlation between the outcomes of the two experiments. While the number of participants who employed two different articulatory strategies to produce the target vowel [ɨ] was 13, only five participants were able to do so with the fricative [sʲ]. The discrepancy between these numbers could be explained by the difference in the numbers of articulatory parameters participants had to adjust during the first and the second experiment. While during the first experiment participants were merely required to change their tongue position in one dimension to compensate for the F2 shifts, compensatory articulations of fricatives potentially required them to adjust three parameters relevant for sibilants production, namely the constriction location, tongue grooving, and the jaw height [16]. The latter articulatory adjustments are less obvious and as a consequence the adaptation process may take longer and less speakers are able to identify the appropriate articulatory adjustments to compensate for the perturbation. Overall, this line of reasoning suggests that the articulatory-to-acoustic mapping in fricatives (and perhaps more generally in consonants) is less transparent compared to vowels as our participants were less successful in identifying the necessary articulatory adjustments to correct for acoustic errors during fricative perturbation than during vowel perturbation.

Our findings indicate that speakers' representations of consonants and vowels indeed differ, however, the nature of this difference is not explicable by the hypothesis that speakers rely more strongly on somatosensory and less on the auditory feedback during the production of consonants since the results of both experiments also clearly demonstrate that auditory feedback is relevant during the production of vowels and fricatives. Our findings rather support the hypothesis that the achievement of the speech production goals is dependent on the "speaker-specific internal representation of articulatory-to-acoustic relationships" [2] but also that the complexity of these relationships is speech sound-specific.

Acknowledgements. We gratefully acknowledge support by DFG grant 220199 to JB. We thank all participants who took part in the study.

References

1. Fowler, C.A., Turvey, M.T.: Immediate compensation in bite-block speech. Phonetica **37**(5–6), 306–326 (1980)
2. Savariaux, C., Perrier, P., Orliaguet, J.P.: Compensation strategies for the perturbation of the rounded vowel [u] using a lip tube: a study of the control space in speech production. J. Acoust. Soc. Am. **98**(5), 2428–2442 (1995)
3. Baum, S.R., McFarland, D.H.: The development of speech adaptation to an artificial palate. J. Acoust. Soc. Am. **102**(4), 2353–2359 (1997)
4. MacDonald, E.N., Goldberg, R., Munhall, K.G.: Compensations in response to real-time formant perturbations of different magnitudes. J. Acoust. Soc. Am. **127**(2), 1059–1068 (2010)
5. Rochet-Capellan, A., Ostry, D.J.: Simultaneous acquisition of multiple auditory–motor transformations in speech. J. Neurosci. **31**(7), 2657–2662 (2011)
6. Lametti, D.R., Nasir, S.M., Ostry, D.J.: Sensory preference in speech production revealed by simultaneous alteration of auditory and somatosensory feedback. J. Neurosci. **32**(27), 9351–9358 (2012)
7. Ghosh, S.S., Matthies, M.L., Maas, E., Hanson, A., Tiede, M., Ménard, L., Perkell, J.S.: An investigation of the relation between sibilant production and somatosensory and auditory acuity. J. Acoust. Soc. Am. **128**(5), 3079–3087 (2010)
8. Villacorta, V.M., Perkell, J.S., Guenther, F.H.: Sensorimotor adaptation to feedback perturbations of vowel acoustics and its relation to perception. J. Acoust. Soc. Am. **122**, 2306–2319 (2007)
9. Brunner, J., Ghosh, S., Hoole, P., Matthies, M., Tiede, M., Perkell, J.S.: The influence of auditory acuity on acoustic variability and the use of motor equivalence during adaptation to a perturbation. J. Speech Lang. Hear. Res. **54**(3), 727–739 (2011)

10. Guenther, F.H., Hampson, M., Johnson, D.: A theoretical investigation of reference frames for the planning of speech movements. Psychol. Rev. **105**(4), 611 (1998)
11. Cai, S., Boucek, M., Ghosh, S.S., Guenther, F.H., Perkell, J.S.: A system for online dynamic perturbation of formant frequencies and results from perturbation of the Mandarin triphthong /iau/. In: Proceedings of the 8th International Seminar on Speech Production, pp. 65–68 (2008)
12. Bolla, K.: A Conspectus of Russian Speech Sounds. Hungarian Academy of Science, Budapest (1981)
13. Bates, D., Mächler, M., Bolker, B., Walker, S.: Fitting linear mixed-effects models using lme4. J. Stat. Softw. **67**(1), 1–48 (2015)
14. Kuznetsova, A., Brockhoff, P.B., Christensen, R.H.B.: lmerTest: Tests in Linear Mixed Effects Models. R package version 2, 0-252015 (2016)
15. Wood, S.N.: Fast stable restricted maximum likelihood and marginal likelihood estimation of semiparametric generalized linear models. J. R. Stat. Soc. Ser. B (Stat. Methodol.) **73**(1), 3–36 (2011)
16. Brunner, J., Hoole, P., Perrier, P.: Adaptation strategies in perturbed /s/. Clin. Linguist. Phonet. **25**(8), 705–724 (2011)

Speech Planning and Comprehension

Investigation of Speech-Planning Mechanism Based on Eye Movement

Jinfeng Huang[1], Di Zhou[1], and Jianwu Dang[1,2(✉)]

[1] Japan Advanced Institute of Science and Technology, 1-1 Asahidai, Nomi,
Ishikawa 923-1292, Japan
{lllhjf, s1610091, jdang}@jaist.ac.jp
[2] Tianjin University, Tianjin, China

Abstract. Eye movements can reflect the brain activities in word recognition and speech planning processes during reading and spontaneous speech comprehension. Most of the previous studies used isolated words alone to investigate the latency time of speech planning. However, it is difficult to explore the mechanism for speech planning in the real situation. In this paper, we used continuous speech to investigate the mechanism of speech planning by means of matching eye movements with the utterances during reading Chinese sentences. The plan units of the reading speech were estimated using the fixation of the eye movements, and the latent times for speech planning were measured for each unit in the reading sentences. It is found that the majority of the planning units was a grammatical word, while most of the four-syllable words were planned as two disyllable units. The latent time of the planning units decreases gradually along the time axis of a sentence. When a sentence consists of two sub-sentences, the decline tendency of the latent time was reset in the boundary between the two sub-sentences. When removing the meaning of the sentence by randomizing the word order, however, the declined tendency of the latent time was disappeared. These phenomena showed that the posterior words of the sentences may save time in semantic comprehension since the preceding words provided more semantic information for posterior ones. We can conclude that the semantic comprehension is an important part for speech planning as well as motor command designs, although comprehension is not an explicit task in the uttering.

Keywords: Speech planning · Eye movements · Latent time
Semantic comprehension · Planning unit

1 Introduction

A major issue in speech production research is how speakers do speech-planning, and what the major factor affects the time of speech planning. Reference with to the process of speech production in Fig. 1, the triggering conditions for executing speech planning can be summarized as two ways. In the case of conversation, a speaker conducts a speech planning based on oneself speech intentions first, and then execute the planned motor commands to control articulatory movements, and finally generate speech sound. In the case of reading, speakers first obtain text information via their eyes, and

© Springer Nature Switzerland AG 2018
Q. Fang et al. (Eds.): ISSP 2017, LNAI 10733, pp. 175–187, 2018.
https://doi.org/10.1007/978-3-030-00126-1_16

comprehend the semantic information, and then carry out the same procedures as those in conversation [1].

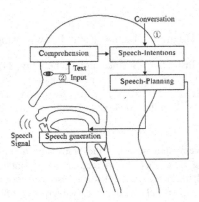

Fig. 1. Schematic of human utterance process

According to the consideration above, some studies investigated the process of speech planning during reading isolated words by measuring the response time from the word presenting to onset of the uttering. They found the frequency and the length of the words would significantly affect the average latent time of speech planning [2–7]. However, the results obtained from the isolated words are difficult to investigate the speech planning for real situation in which utterance is continuous but not isolated.

Some other studies combined the eye movement with acoustic measure, where they believed that eye movements can reflect speech planning processes during reading [6, 8–10]. Meyer et al. (1998) used the time difference between the onset of the eye leaving a word and the onset of uttering the word as the response latency [6]. In uttering a sentence continuously, the flows of obtaining the text word by word, planning and uttering the words are considered to be executed simultaneously. There is no evidence to show that the onset of the eye leaving a word is the onset of planning. Furthermore, we found that in the latter part of a sentence the onset of uttering a word is earlier than the onset of the eye leaving the word. In the criterion of Meyer et al. (1998), the latent time will be a negative value. It is not reasonable. For this reason, we define the period from the onset of eye reaching a word to the onset of uttering the word as "the latent time", which was used in the latent time measurement by acoustics only. According to Rayner (1998), three standard eye movement parameters are selected and used in this study, which are the first fixation if there are multiple fixations on a target word, mean fixation duration that is the average of all effective fixations in a sentence, and total viewing time that is the summation of all fixations duration on the sentence [9].

In previous research, they only investigated speech planning in isolate words situation. It was hard to investigate the speech planning for real situation which is continuous sentence but not isolated. To bridge this gap, we designed a series of experiments to discover the mechanism of speech planning in continuous speech. This paper is organized as the following. Section 1 is introduction, Sect. 2 is about the experiment detail and data analysis. Section 3 is our result discussion and Sect. 4 is conclusion about our research.

2   Experiments and Data Analysis

To investigate the mechanism of speech planning, we designed three experiments. In Experiment 1, we used simple sentences which contains only 11 disyllabic words. In Experiment 2, we used a complex sentence which consists of words with different syllables to investigate what the units are used in speech planning. In Experiment 3, we used the nonsense sentence with random word in order to investigate whether the semantic comprehension affects the latent time or not.

2.1   Experiment 1

In the past research, most of them used isolated words or carry sentence in the investigations of the mechanism of speech planning, which is different from the situation of continuous speech. For this reason, we designed an experiment to investigate how the latent time of speech planning changes in the continuous speech.

Reading Materials
In order to exclude the effects of the word length on the latent time, we construct ten sentences using the disyllabic words alone, with clear grammatical boundaries. To meet this requirement, we choose Chinese as the materials in which one character is only pronounced one syllable. Each of the ten sentences contains 11 disyllabic words. All of the 11 disyllabic words were high-frequency words. The materials are presented in one line on the center of the screen sentence by sentence.

Participants
Ten graduate students of JAIST (6 male and 4 female) served in the experiment as participants, whose native language is Mandarin. Their age ranged from 23 to 28 years old. They have normal or corrected-to-normal vision via contact lenses or glasses. All participants were inspected through the eye movement and pronunciation testing to make sure that the participates have the same level of understanding the reading materials and in the background knowledge.

Experimental Procedure
The experiment process was controlled by one host computer. When the subjects read the presented sentence on the screen, their eye movements were monitored and recorded via an eye tracking system (Eyelink 1000 Plus, SR Research Ltd, Ontario, Canada), the sampling rate was 1000 Hz for eye movement. Speech was recorded simultaneously using a microphone (SONY ECM MS957), the sampling rate was 44100 Hz. In the experiment, the computer recorded the onset and offset times of the fixation on a target word, and also spatial coordinates of the participants' fixations. Both eyes were monitored, the data used in this study were the average of left and right eyes' movements (Fig. 2).

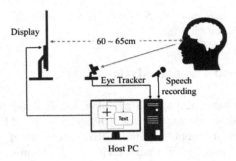

Fig. 2. Setup of the Experiments.

To carry out this experiment, we designed a system of eye movement and speech measurement by using MATLAB and PsychToolBox (version: 3.0.14). Participants learnt the experiment procedure by reading experiment instruction. During the experiment, participants placed their foreheads against a rest to prevent movements in depth and to keep their eyes about 65 cm from the surface of the monitor. We calibrated the system using the validation routine by a nine-point calibration, and made sure the deviation of Gaze Accuracy less than 0.50°. As seen in Fig. 3, each trial started with a fixed signal '+' presented for about 1000 ms to attract subjects' attention. In order to eliminate the delay when the microphone and eye trackers start recording, it advanced start recording after the '+' mark appeared for 400 ms and 700 ms. When a sentence was presented on the screen, we asked the subject to utter the sentence as quickly as possible, and keep a normal speech rate. During the uttering, eye movement and speech sound were recorded simultaneously. When a participant finished reading of a sentence, they were required to end the recording by pressing the button 'Esc'. After the button is pushed, the recording would stop in 300 ms after the button is pushed. The experimental procedure is shown in Fig. 3.

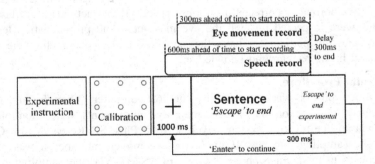

Fig. 3. Eye movement and Speech measurement system.

Data Analysis and Results

In analyzing the eye movements offline, the locations and durations of the participant's gaze points were superimposed onto the sentence as spot in each trial via MATLAB

Fig. 4. The fixation on a sentence.

software, where the longer the duration and the larger the area of the spot, as shown in Fig. 4.

Figure 5 plots the trajectory of the eye movement and spectrogram of speech sound with word boundaries for a given sentence. Horizontal axis represents the time index of the continuous speech with its spectrogram as a background, while the vertical axes show the word boundary with reference to eye movements. The eye movement is plotted along two time-axis. One can intuitively observe and verify how the gaze points are distributed on the words and how long the durations are, and also can find out the onset and offset of the fixation on each word. The onset of the fixation is defined as the start timing of speech planning for this word. For the reading sound, we used a force alignment method to get the word boundary in the continuous speech [11]. As shown in Fig. 5, when subject pronounced the first-word "今年" (this year), the gazing position already reached the second-word "除夕" (New Year's Eve), and it moved to the third word at the latter part of the pronunciation of the first word. One can see that the uttering has a large delay comparing with the eye movement, where the delay responds to the latent time of speech planning.

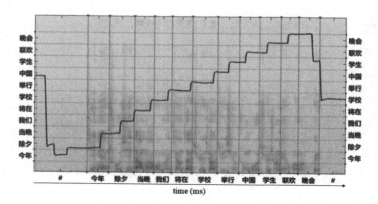

Fig. 5. The relation between eye movement and speech.

We calculated the time difference between the onset of eye reaching a word and the onset of uttering the word for each word. The difference is defined as "the latent time". Table 1 shows the results of the average latent time of the first word of ten sentences over all subjects, and the word number of the gazing location at the onset of uttering the first word. The latent time of the first word is about 550 ms (AVG = 558 ms, STD = 240). It is a little bit longer than that of the isolated word case, in which the latent time was about 467 ms [4].

Table 1. Average of LT and GP for ten subjects. (LT: Latent Time of the first word; GP: word number of the Gazing Position; AVG: average; STD: Standard Deviation.)

	LT (STD)	GP
AVG	558 (240)	2

The same analysis approach was applied to each word in the sentences, and thus the latent time was obtained for each word over all sentences and all subjects. Figure 6 shows the latent time distribution of 11 disyllable words in 10 sentences for ten subjects, where the dashed lines are the linear-regressions of the latent time. One can see that the latent time decreases gradually along with the time axis, and the time difference of the latent time between the first word and the last word reaches 360 ms (AVG = 363 ms, STD = 220), although all of the word lengths were the same. The latent time were analyzed using analysis of variance (ANOVA). The latent time of the first word and the last word had significantly different (F(1 ,18) = 16.76, p < 0.001). It indicates that the latent time of speech planning changes with the location of the words in the continuous speech.

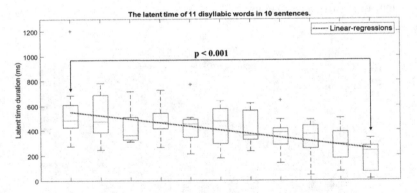

Fig. 6. Latent time trend from disyllable sentences.

Rayner (1998) measured the duration of the fixation in silent and oral reading for a number of languages, and found that the average fixation duration was 200 ms or more. In our experiment, the average fixation duration is 256 ms with a standard deviation of 134, which is consistent with the observation of Rayner (1998). We analyzed all fixation within normal sentence between the first fixation of the first word and the last fixation of the last word. As the result, the number of fixation points almost consistent with the number of disyllable grammatical words in the sentence (AVG = 11.57 fixation points, STD = 0.52) that was 11 disyllable words for each sentence. One disyllabic word has about one fixation (AVG = 1.05 fixation, STD = 0.05). It demonstrates that the period of 200 ms is reasonable to deal with the planning unit. Accordingly, we suppose the planning unit in speech production have a fixation of 200 ms and more, so that delete the fixation with a duration less than 200 ms for clarifying the units used in speech planning in the following analysis.

2.2   Experiment 2

From Experiment 1, we found that latent time decreases gradually as the word location moves to the end of the sentence, and the units of speech planning is consistent with the grammatical word, while all of the sentences consist of disyllabic words. If the word length varies, both the latent time and the planning units may be changed. Another interesting issue is that how to the latent time change when elongating the sentence sufficiently. For these reasons, we investigate the planning mechanism using complex sentences including variety word lengths.

Reading Materials

We constructed 22 Chinese sentences for Experiment 2. Each sentence contains the words with different syllables, whose number was ranged from one syllable to four syllables. All of the words have high familiarity. Among the reading materials, there are seven simple sentences and 15 complex sentences, where the complex sentences consist of two sub-sentences.

Participants

Twenty graduate students were recruited from JAIST (15 male and 5 female) for this experiment, whose native language is Mandarin. Their age ranged from 22 to 29. Ten of them had taken part in Experiment 1.

Experiment Procedure

The eye tracker (Eyelink 1000 Plus, SR Research Ltd, Ontario, Canada) recorded eye movement with a sampling rate of 1000 Hz, and microphone (Rode NTG-3, RØDE Microphones LLC) recorded speech sound with a sampling rate of 44100 Hz, simultaneously. The onset and offset times, as well as spatial coordinates of the participants' fixations were measured in this experiment. Both eyes were monitored, the data used in this study are the average of left and right eyes. The experimental procedure is same to the Experiment 1 shows in Fig. 3.

Analysis and Results

In Experiment 2, we used the same analysis method as Experiment 1 to measure the latent time and estimate planning units. Figure 7 shows the latent time distribution of each words in 7 sentences for 20 subjects, where the dashed lines are the linear-regressions of the latent time. The analysis method was the same as above. The latent time of the first word and last word had significantly different ($F(1, 36) = 31.81$, $p < 0.001$). The result demonstrates that the same tendency of the latent time as that of Experiment 1, even if the words have quite different syllable number.

Figure 8 shows the distribution of the latent time measured from the complex sentences, which consists of two sub-sentences. In the figure, the vertical black line was the boundary of the two sub-sentences. The results show that the tendency of the latent

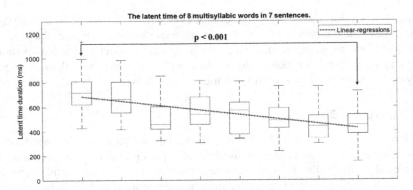

Fig. 7. Latent time trend from multisyllabic sentences.

times looks like to be reset in the boundary between the sub-sentences, but the decline tendency did not change within each sub-sentence. In the first sub-sentence, the latent time of the first word is significantly different from the last word ($F(1, 36) = 43.97$, $p < 0.001$). The latent time of the second sub-sentence shows the same tendency of, where the latent time of the first word is significantly different from the last word ($F(1, 36) = 36.3$, $p < 0.001$).

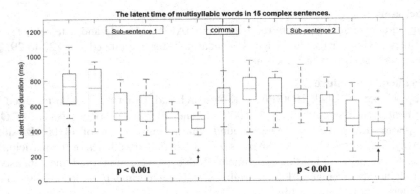

Fig. 8. Latent time trend from two sub-sentences.

For analyzing the planning units used in the speech planning, we first counted the appearance frequency of the grammatical words with different length in our corpus. The appearance frequency is shown in the left penal of Fig. 9, where the horizontal axis represents the syllable number within a grammatical word, and the vertical axes show the appearance frequency of the grammatical words. This distribution reflects the ratio of the words with different syllables in the corpus. In the observation, we counted the pairs of a fixation and its corresponding syllables, and then cluster the pairs according to the syllables per fixation. The distribution of the syllable number per fixation is shown in the right panel of Fig. 9.

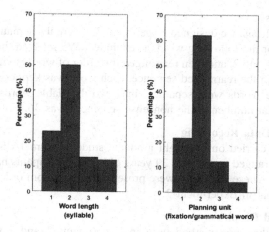

Fig. 9. The appearance frequency of the real word length and the planning units.

In the right panel, one can see that the appearance frequency of monosyllabic grammatical words is about 20%, which is almost the same as the frequency in the corpus. That indicates that a one-syllable word is usually treated as an individual planning unit. Similarly, three-syllable grammatical words in the right panel have the same percentage, about 10%, as that in the left panel. The disyllables have higher percentage than that in the corpus, while the four-syllable words have lower percentage than that in the left panel. This phenomenon implies that a part of the four-syllable words may be separated into two disyllabic words in the planning stage. In order to clarify the details, we analyzed four-syllable words and found about two-thirds of them have two fixations point (AVG = 2.31 fixation, STD = 0.59). The results indicate that the unit of speech planning for Chinese language is basically the grammatical words when the word has three or less syllables. However, most of the four-syllable grammatical words were separated into two disyllable words in the speech planning.

2.3 Experiment 3

The results of Experiments 1 and 2 showed a general tendency that the latent time of speech planning gets shorter if the words are located in the posterior part of a sentence. For complex sentences, the latent time was reset in the boundary of two sub-sentence, while the tendency in each sub-sentence is the same as that in simple sentences. In the continuous speech, the effects of word length and word frequency on the latent time are not so significant as word locations. In average, the last word of the sentence is about half of the latent time comparing with the first word.

Why is the word location so important for the latent time? We speculate that the latent time may consists of two components: semantic comprehension and motor command design. If so, it means that the semantic comprehension time would be shorter for the posterior words because the preceding words provided more semantic information for the follows.

Reading Materials

To prove the hypothesis, we designed Experiment 3, where the semantic support of the preceding words for the following words is excluded. We selected the same sentences from Experiments 1 and 2, and then rearranged the order of words to make the sentence have no meaning. In the rearranged sentence, each word was kept correctly, except all of the four-syllable words were separated into two disyllable words before the rearrangement. The total number of the nonsense sentences was 36.

Participants and Data Recording

Experiment 3 was carried out on eight graduate students from JAIST (5 male and 3 female). Their age ranged from 24 to 29 years old. The participants had taken a part in Experiment 1 or 2. These sentences were presented in a random order. Data recording was the same as that in Experiment 2.

Data Analysis and Results

In Experiment 3, the same method used in Experiments 1 and 2 was employed to analyze the eye movements and speech sound. Figure 10 shows the results obtained from the nonsense sentences. In the nonsense sentences, the latent time of first word and last word has no significantly different. For the disyllabic words sentence, the ANOVA result is $F(1, 140) = 1.36$ and $p = 0.25 > 0.1$. For the multisyllabic words sentence, the ANOVA result is $F(1, 172) = 0.15$ and $p = 0.699 > 0.1$. In the complex sentences, the result is also same in each sub-sentence ($F(1, 158) = 1.51$, $p = 0.22 > 0.1$ and $F(1, 158) = 1.57$, $p = 0.24 > 0.1$). One can see that unlike the normal sentence, the latent time does not change with the word location since there were no the semantic effects in the sentence level. The latent time of each word about 650 ms (AVG = 659 ms, STD = 229) which is consistent with the result of the isolated words [7]. However, the variation is larger than that observed in the normal sentence. The reason is that some subject tried to retrieve some semantic meaning of the sentences during the uttering.

3   Discussion

In this paper, we investigated the mechanism of speech planning in continuous speech by combining eye movement and speech sound. The mechanism was estimated by the variations of the latent time and the basic plan unit during speech planning.

Most of the previous studies used isolated words to measure the latent time of speech planning, the general conclusion based on those studies was that the latent time was related to the word length. The longer the words and the longer the latent time [3, 5]. In this study, we used different types of continuous utterances to verify the previous conclusion. As the results, the latent time decreases gradually along with the time axis. For the sentences with two sub-sentences, the decline tendency was consistent within each sub-sentence, while the latent time was reset in the boundary of two sub-sentences. The latent time linearly reduced as the word is located in the posterior part of the sentences. These phenomena demonstrated that the latent time of speech planning is heavily dependent on the word's location in the continuous speech, but not on their

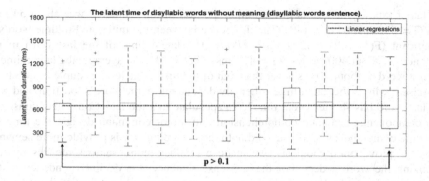

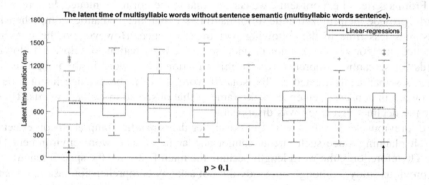

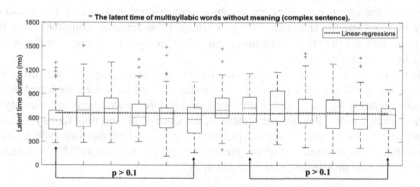

Fig. 10. Latent time obtained from sentences without meaning.

word length. Accordingly, our results do not support the previous conclusion that the latent time of speech planning was proportional to the word length.

In our study, in the cases with normal (meaningful) sentences, the latent time of the initial word was about 700 ms (AVG = 713 ms, STD = 195) for speech planning, and the latent time of the words in the middle portion of the sentences was about 540 ms (AVG = 542 ms, STD = 149). The difference between the initial and middle words is significant (F(1, 132) = 32.86, p < 0.001). The latent time of the last word in the sentences is about 400 ms (AVG = 412 ms, STD = 147). In average, the latent time of the last word is about 40% shorter than that of the first one. Here, a couple of questions are raised: why is the latent time is heavily dependent on the word's location, and why the last word has shorter latent time? We speculate that the latent time may consist of two components: semantic comprehension and motor command design, the time for semantic comprehension is reduced due to the preceding words provide more semantic cues. To prove our hypothesis, we made a number of nonsense sentences by randomizing the word orders. For the nonsense sentence, the declined tendency of the latent time obviously disappeared. In average, the invariant latent time is about 650 ms (AVG = 650 ms, STD = 250), which is the same as that observed in isolated words.

From a series of experiments, we can say that in the normal sentence, the preceding words provide helpful semantic information, so that the semantic comprehension consumes less time in the following part of a sentence. However, in the nonsense sentence, previous words cannot provide semantic information to help the following words for semantic comprehension. In this situation, each word in the nonsense sentence is semantically treated as the isolated word. In fact, the average latent time of each word in nonsense sentences is the same as that observed in isolated words. That is why the declined tendency was disappeared.

In previous studies, they could not discover the semantic comprehension effect on speech planning since such effects cannot appear in isolated word environment [12, 13]. Therefore, only the word length became the main factor for the speech planning in the previous study. Although our study did not ask any comprehension task, the results imply that semantic comprehension automatically takes place in the uttering task, and more 40% time was used for the comprehension if no semantic environment. It confirmed that the effects caused by the semantic comprehension is indeed larger than that cause by word length. That means the processing of the semantic comprehension is an important part for the speech planning, and the processing of comprehension is automatic in the sentence level.

For analyses the planning unit that is used in speech planning, we analyze all fixation within the normal sentence. In the disyllabic word sentences, we found that the disyllabic word is a basic planning unit because disyllabic word seems to have only one fixation. In the multisyllabic word sentences, the words with less than four syllables are treated as a planning unit in speech planning, while most of the four-syllable words almost have two fixations or more, therefore most of the four-syllable words were divided into two disyllables in speech planning.

4   Conclusions

Differing from the past studies with isolated words, continuously reading speech was used to investigate the mechanism of speech planning in this study. It is found that the latent time of speech planning in continuous speech decreases gradually as the preceding words increase. To explain the decline tendency of the latency, we assume that the latent time of speech planning consists of two parts: semantic comprehension and motor command design and conducted an experiment to confirm the assumption. It is also found that the semantic comprehension mainly takes place in sentence level rather than in word level. The experiment results indicate that for Chinese the grammatical word is the basic unit for speech planning, while most of the four-syllable words were divided as two disyllables in speech planning. Further investigation of the speech planning mechanism with semantic comprehension and motor command construction is remaining for the future study.

Acknowledgements. The research is supported partially by the National Basic Research Program of China (No. 2013CB329301), and the National Natural Science Foundation of China (No. 61303109). The study is supported partially by JSPS KAKENHI Grant (16K00297).

References

1. Dang, J.: Present state and prospect of speech synthesis technology (review). In: The Evolution of Human and Machine Communication, pp. 125–140 (2015). (Co., Ltd. Nikkeiinsatu ISBN 978-4-86469-065-2) (In Japanese)
2. Kirov, C., Wilson, C.: Bayesian speech production: evidence from latency and hyperarticulation. In: CogSci (2013)
3. Meyer, A.S., Belke, E., Häcker, C., Mortensen, L.: Use of word length information in utterance planning. J. Mem. Lang. **57**(2), 210–231 (2007)
4. Chiu, C.C.: Phonological words in Mandarin speech production. In: Annual Meetin5 of the Berkeley Linguistics Society, vol. 31, no. 1, pp. 61–72), June 2005
5. Meyer, A.S., Roelofs, A., Levelt, W.J.: Word length effects in object naming: the role of a response criterion. J. Mem. Lang. **48**(1), 131–147 (2003)
6. Meyer, A.S., Sleiderink, A.M., Levelt, W.J.: Viewing and naming objects: eye movements during noun phrase production. Cognition **66**(2), B25–B33 (1998)
7. Levelt, W.J., Roelofs, A., Meyer, A.S.: A theory of lexical access in speech production. Behav. Brain Sci. **22**(1), 1–38 (1999)
8. Allum, P.H., Wheeldon, L.R.: Planning scope in spoken sentence production: the role of grammatical units. J. Exp. Psychol. Learn. Mem. Cogn. **33**(4), 791 (2007)
9. Rayner, K.: Eye movements in reading and information processing: 20 years of research. Psychol. Bull. **124**, 372–422 (1998)
10. Richardson, D.C., Dale, R.: Looking to understand: the coupling between speakers' and listeners' eye movements and its relationship to discourse comprehension. Cogn. Sci. **29**(6), 1045–1060 (2005)
11. Yuan, J., Liberman, M.: Investigating consonant reduction in Mandarin Chinese with improved forced alignment. In: Interspeech, pp. 2675–2678 (2015)
12. Guoli, Y., Lanlan, Z., Xia, Z., Shasha, S.: The processing of psychological word in chinese reading. Stud. Psychol. Behav. **3**, 006 (2012)
13. Wei, W., Li, X., Pollatsek, A.: Word properties of a fixated region affect outgoing saccade length in Chinese reading. Vis. Res. **80**, 1–6 (2013)

Global Monitoring of Dynamic Functional Interactions in the Brain During Chinese Verbs Perception

Yuke Si[1], Jianwu Dang[1,2(✉)], Gaoyan Zhang[1(✉)],
and Longbiao Wang[1]

[1] Tianjin Key Laboratory of Cognitive Computing and Application,
Tianjin University, Tianjin, China
jdang@jaist.ac.jp, zhanggaoyan@tju.edu.cn
[2] Japan Advanced Institute of Science and Technology, Nomi, Ishikawa, Japan

Abstract. Previous studies suggested that during speech perception and processing, auditory analyses clearly took place in the auditory cortices in the temporal lobes bilaterally, semantic processing was supported by a temporo-frontal network which strongly lateralized to the left hemisphere while a prosodic processing network mainly located in right hemisphere. However, some studies proposed that the linguistic abilities such as phonology and semantics recruited regions in both hemispheres. To understand the neural mechanism underlying speech perception and processing, it is important to uncover the dynamic functional interaction in the brain. Our aim is to investigate whether a prevalent human brain network exists during perceiving Chinese verbs and how the effective connectivity changes in spatial and temporal domains. An auditory listening experiment was carried out to monitor the brain activations in full-time scale by using the electroencephalograph (EEG) signals recording when native subjects perceiving Mandarin verbs. By performing a Granger causality analysis and statistical analysis, six connection patterns in different time periods were constructed under global monitoring of dynamic brain activities. The results showed different connections and inter-regional information flows in six time intervals. It can be indicated from this study that the bilateral hemispheres not only involves in the speech perception and processing, but also have information interaction. The results are essential for the more detailed bilateral brain network analysis with full-time monitoring in future studies.

Keywords: Speech perception · Dynamic functional interaction
EEG · Source localization

1 Introduction

The neural mechanism of the brain is generally investigated from two aspects: functional separation and functional integration [1–3]. Functional separation investigates what specific functions the given brain region has, while functional integration examines how the multiple brain regions integrate and work for a specific function.

© Springer Nature Switzerland AG 2018
Q. Fang et al. (Eds.): ISSP 2017, LNAI 10733, pp. 188–197, 2018.
https://doi.org/10.1007/978-3-030-00126-1_17

Generally speaking, in speech production and perception, there are a number of brain areas work integratedly but individually.

Previous studies based on functional integration had provided many evidence on the interaction of multiple brain regions during speech perception. Hickok and Poeppel proposed the dual-stream model of speech perception and processing in the human brain and gave a perspective on the information flows among the involved brain regions in the model [4–6]. This dual stream model also indicates a left dominant organization worked for the phonological and semantic representations [7]. Researchers also analyzed the anatomical connectivity of dual-stream model and suggested its role in language processing [8]. In addition, our previous researches as well demonstrated different activation patterns in the left hemisphere of the human brain during people perceiving Chinese words and pseudo words [9], suggesting involved different brain regions and the time duration during perceiving auditory stimuli.

While for the right hemisphere, there was a reportedly prosodic processing model including an auditory ventral pathway along the superior temporal lobe, while an auditory-motor dorsal pathways connecting posterior temporal and inferior frontal/premotor areas worked for rhythmic and melodic variations in speech information [10, 11]. However, in the dual-stream model of speech perception and processing, the author also suggested that there was some degree of bilateral capability in lexical and semantic access, which means that both bilateral hemisphere may be recruited to accomplish the lexical and semantic processing. This view point was approved by some researches. The author in [12] indicated that during words perception, the left hemisphere was activated to process the most strongly related meanings and the right hemisphere worked for other broader meanings including subordinate and complex meanings. In addition, the functional MRI (fMRI) study suggested that both bilateral hemispheres played an important role in the lexical-semantic information processing, but their functions and roles may be different [13].

The aforementioned studies showed more and more evidence suggest the involvement of both hemisphere during speech processing, while the functional interaction between the left hemisphere and the right hemisphere, especially during different temporal stages is still not clear. In order to understand the neural mechanism underlying speech perception and processing, it is necessary to explore full-time spatiotemporal functional connections and integration patterns. In this study, our aim is to investigate both temporal and spatial properties of dynamic global brain network and its information flows during perceiving Chinese verbs by using a high density EEG technology, series of data preprocessing methods, source localization analysis and a full-time dynamic granger causality analysis, also clarify the relation between the left hemisphere and the right hemisphere of human brain in the lexical-semantic network and their specific roles during words perception.

2 Experiment and Method

2.1 Participants

Sixteen healthy subjects from Tianjin University (8 males, 8 females) aged from 21 years to 28 years (Mean age is 23.9 years. Standard deviation is 1.59 years.) were recruited for this experiment. All the subjects were right-handed native Mandarin speakers with normal vision and hearing, and without any history of psychiatric disorders or neurological deficit. The study was conducted in accordance with the Declaration of Helsinki [14]. This study was approved by the local ethics committee. All the subjects gave their written informed consent and were paid for their participation.

2.2 Materials

The materials consisted of 50 Chinese disyllabic verbs and 50 Chinese disyllabic pseudo words with both auditory and visual stimuli. The auditory stimuli were recorded by a Mandarin male announcer in a single channel of 16 bit resolution and sampled at 44.1 kHz with a duration of 900 ms. The visual stimuli were semantically congruent or incongruent disyllabic verbs in Chinese.

Besides, verbs had the mean frequency of occurrence of more than 14 per million [15] and the average occurrence of one hundred words was 21.14 per million. Furthermore, strokes of the Chinese meaningful words had a mean number of 18.03 and the Chinese pseudo words were 18.34.

2.3 Experimental Procedure

The subjects were instructed to accomplish tasks in an electronically and magnetically shielded soundproof room.

The experimental procedure is shown in Fig. 1. At the beginning, the trial started with a white cross appeared in the center of the PC monitor with a black background to draw the subject's attention, which lasted for a random duration of 200 ms to 300 ms. Next the subject heard a 1200 ms auditory stimuli through a headphone set. After a 400 ms interval, the Chinese words contain two Chinese characters were displayed for 600 ms on the screen. Then the subject was asked to judge the consistency of the auditory and visual stimulus. This task is used to guarantee that the subject was concentrated and able to distinguish different words categories during EEG signal recording. The interval between two trials was an undetermined duration of 800 ms to 1000 ms. There were four conditions which combined two types of words and two coincidences of audio-visual stimulus. In condition 1, the same meaningful words were presented in audio and visual parts, while in condition 2, the different meaningful words were shown in the two parts. Analogously, the same pseudo words were displayed in condition 3 while different ones in condition 4.

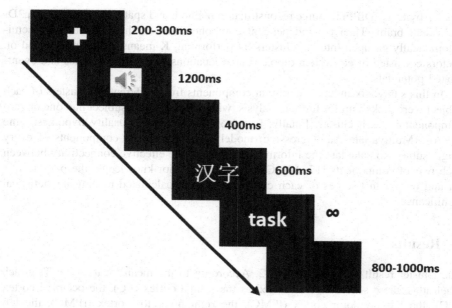

Fig. 1. Experimental procedure.

2.4 EEG Signals Recording and Data Analysis

The EEG signal was recorded through the Neuroscan Synamps system (Neuroscan, USA) with a 128 electrodes cap at a sampling rate of 1000 Hz. Six channels were used for recording vertical electrooculogram (VEOG) and horizontal electrooculogram (HEOG). The impedance of each electrode was kept below 5 kilo-ohm during the data acquisition.

This study analyzed the EEG data only recorded for the auditory stimulus of real Chinese verbs. Two subjects' data were discarded because the electrode detached during EEG signals recoding. Mean response accuracy of 99.8% ± 0.1% indicated that participants focused on the whole tasks.

The raw EEG data were processed using EEGLAB toolbox [16] in MATLAB (Mathworks Inc., Natick, MA). The data analysis procedure can be divided into two parts, which include data preprocessing and functional connectivity analysis. In the preprocessing part, the data were firstly filtered by a low pass filter with cut-off-frequency of 45 Hz to remove high-frequency electrical artifacts and then down sampled to 250 Hz. Secondly, the data were filtered using a high-pass filter with 1 Hz to remove the baseline drift. After that, the data of bad channels and line noise was removed. Then the data were re-referenced to average.

After preprocessing, the brain source information flow analysis based on Granger causality hypothesis was performed on the EEG data. Firstly, the independent component analysis (ICA) was conducted after data rank adjustment to separate instantaneously effective independent brain processes from physiology and artificial noise. Secondly, all these effective brain sources were labeled as single equivalent current dipoles by using DIPFIT source reconstruction method, and spatially localized in a 3D-equivalent brain. Then the independent components of all the subjects were semi-automatically grouped into six clusters by performing K-means clustering method on vectors consisted of equivalent dipole source locations, scalp topographies and event-related potentials.

In this study, six independent brain components from different six clusters of each subject were picked up for further analysis, which means every subject had one or zero component in each cluster. Finally, based on the Granger causality hypothesis, the MVAR (Multivariate Autoregressive) model was built on the components of every single subject to calculate the information flow and the effective connectivity between each pair of components [17, 18]. Combining with prior knowledge, the p-value was set and the mean t-scores of each connectivity were calculated to indicate statistical significance.

3   Results

The Cluster result is shown in Fig. 2. According to the mean locations in Talairach label atlas, these sources were labeled as the frontal cortex (FC), the occipital cortex (OC), the left premotor cortex (lPMC), the right premotor cortex (rPMC), the left temporal cortex (lTC), and the right temporal cortex (rTC).

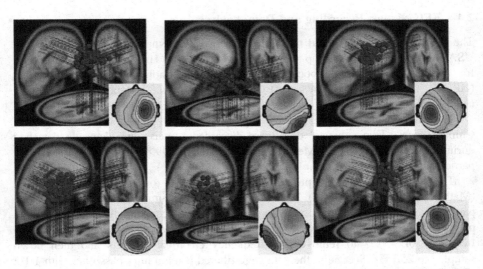

Fig. 2. mean scalp topographies of six clusters. From upper-left to lower-right are rPMC, rTC, lPMC, OC, lTC, FC respectively.

With analyzing the EEG source information flow, we examined the individual variations of effective connectivity between independent components in different brain regions. Detailed six brain areas refer to the left inferior frontal gyrus in FC, the left lingual gyrus in OC, left precentral gyrus in lPMC, left superior temporal gyrus in LTC, right supplementary motor area in rPMC, and right middle temporal gyrus in rTC. According to the spacial and temporal features and differences of changes throughout the process, six important time periods was found in the results, which reflected the characteristics of the brain activity during perceiving verbs in different stages. The significant effective connections were plotted in Fig. 3. Here we set the p-value to 0.1 (combined with prior knowledge) and then calculated the mean t-score of each effective connectivity in whole time duration to represent its specific statistical significance. The result is shown in Table 1. Note that a brain region which had no information interaction with other brain regions did not indicate that the region was not activated.

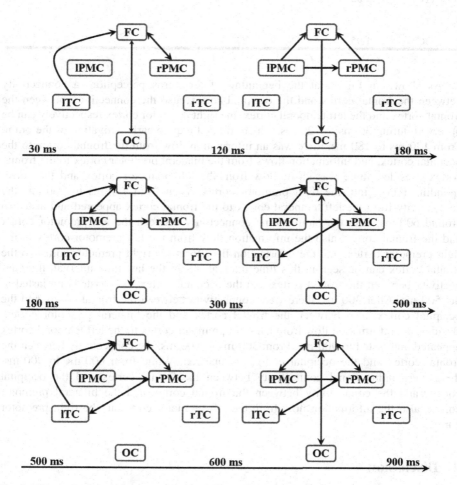

Fig. 3. Dynamic functional networks in different time intervals during verbs perception and processing.

Table 1. Mean t-score of each effective connectivity in whole time duration.

↗	FC	lPMC	rPMC	lTC	rTC	OC
FC		1.71	2.11			2.24
lPMC	1.71		2.18			
rPMC	2.27			1.79		
lTC	2.05					
rTC						
OC	1.95					

As shown in Fig. 3, at the beginning of the verbs perception, a connectivity between the frontal cortex and the occipital cortex, also the connectivity between the frontal cortex and the left temporal cortex, the right premotor cortex respectively can be observed during 30 ms to 120 ms, which implied a quite early activation of the brain. From 120 ms to 180 ms, there was an information flow from the frontal cortex to the occipital cortex, and information flows from the bilateral premotor cortex to the frontal cortex. Besides, there was an outflow from the left premotor cortex and the corresponding inflow into the right premotor cortex. Then from 180 ms to 300 ms, the connectivity from the left temporal cortex to the frontal cortex appeared and lasted to around 600 ms. And the bidirectional connectivity between the left premotor cortex and the frontal cortex, also the information flow from the left premotor cortex to the right premotor cortex, and the information flow form the right premotor cortex to the frontal cortex can be seen in this time interval. As in the last time interval, the connectivity between the frontal cortex and the occipital cortex still existed and lasted to the 500 ms. After 300 ms, there were connectivity between the frontal cortex and the occipital cortex, also between the frontal cortex and the bilateral premotor cortex. Besides, the information flow from the right premotor cortex to the left temporal cortex appeared and lasted to 900 ms. From 500 ms to 600 ms, the connectivity between the frontal cortex and the occipital cortex disappeared. While from 600 ms to 900 ms, there were bidirectional connectivity between the frontal cortex and the occipital cortex, and the connectivity between the frontal cortex and the bilateral premotor cortex, also the information flow from the left premotor cortex to the right premotor cortex.

4 Discussion

In this study, we used dynamic granger causality analysis based on EEG source reconstruction to investigate the global brain network during speech perception and processing, and revealed the specific functional interactions in the whole brain during different temporal stages.

As for the left hemisphere, there are two significant connectivity, one of which is between the left temporal cortex and the frontal cortex, and the other one is between left premotor cortex and other brain regions. On the one hand, it can be seen that between 180 ms and 600 ms, the frontal areas had information flow with the left temporal cortex. Here the FC was specifically referred to the inferior frontal regions. This results are consistent with some studies [4, 13, 19, 20] and functionally referred to phonological and lexical-semantic processing [13]. In our previous results [9], we also found that the connection between the frontal areas and the left temporal areas appeared in the interaction stage of dual-stream model during speech perception and processing, which can be regarded as the characteristic connection during the interaction of the two streams. In addition, this connection appeared after 180 ms, which is consistent with the P200 components in some ERP studies representing the important temporal features in lexical processing [21]. On the other hand, the involvement of the left premotor cortex may be due to the reflection from auditory sound to articulation and the semantic action representation in verbs [22–27].

For the right hemisphere, our results showed the information flow from and to the right premotor cortex, while there was no connectivity between the right temporal cortex and the other brain regions. Since previous studies demonstrated that both left and right temporal cortex involved in semantic and lexical processing [12, 28]. Here our results failed to support the information flow from and to the right temporal cortex. That may be due to the right temporal cortex involved in the speech perception and processing in a way of activation instead of interacting with other brain regions. In terms of right premotor cortex, there were researchers state that the activation of the bilateral motor cortex both related to action-related verbs processing [12, 24]. Our results also supported that the left and the right motor system not only involved in the Mandarin verbs perception by interacting with other related brain regions, but also have information exchange with each other.

5 Conclusion

Generally, the above results illustrate the dynamic brain connectivity variation during speech perception by taking advantage of the high temporal resolution of EEG and series of data processing methods. The results show that the bilateral hemispheres not only involve in the speech perception and processing, but also have information exchange, which is essential for the more detailed bilateral brain functional connectivity analysis with full-time monitoring in future studies. Although previous EEG data processing methods can be applied to give a perspective of brain activation in some time points or brain regions, there still lack more comprehensive results and high temporal resolution methods to clarify full-time authentic global brain activity. This dynamic brain network, which can be monitored both spatially and temporally at the same time, can provide more evidence to present neuromechanism research and theory on speech perception and processing, as well as a different viewpoint to understand and explain some inconsistent results in different current studies.

Acknowledgements. The research is supported partially by the National Natural Science Foundation of China (No. 61503278 and No. 61771333). The study is supported partially by JSPS KAKENHI Grant (16K00297). Besides, we express sincere gratitude to Makoto Miyakoshi for providing support in this study.

References

1. Lyu, B., Ge, J., Niu, Z., Tan, L.H., Gao, J.H.: Predictive brain mechanisms in sound-to-meaning mapping during speech processing. J. Neurosci. **36**(42), 10813 (2016)
2. Vinodh, K.G., Tamesh, H., Jaiswal, A.K., et al.: Large scale functional brain networks underlying temporal integration of audio-visual speech perception: an EEG study. Front. Psychol. **7**, 1558 (2016)
3. Herrmann, B., Maess, B., Hasting, A.S., et al.: Localization of the syntactic mismatch negativity in the temporal cortex: an MEG study. Neuroimage **48**(3), 590–600 (2009)
4. Hickok, G., Poeppel, D.: The cortical organization of speech processing. Nat. Rev. Neurosci. **8**(5), 393–402 (2007)
5. Hickok, G., Poeppel, D.: Dorsal and ventral streams: a framework for understanding aspects of the functional anatomy of language. Cognition **92**(1), 67–99 (2004)
6. Ahveninen, J., et al.: Task-modulated "what" and "where" pathways in human auditory cortex. Proc. Natl Acad. Sci. U. S. A. **103**(39), 14608 (2006)
7. Rauschecker, J.P., Scott, S.K.: Maps and streams in the auditory cortex: nonhuman primates illuminate human speech pro-cessing. Nat. Neurosci. **12**(6), 718 (2009)
8. Sarubbo, S., De Benedictis, A., Merler, S.: Structural and functional integration between dorsal and ventral language streams as revealed by blunt dissection and direct electrical stimulation. Hum. Brain Mapp. **37**(11), 3858–3872 (2016)
9. Si, Y., Dang, J., Zhang, G.: Investigation of the spatiotemporal dynamics of the brain during perceiving words. In: 2016 10th International Symposium on Chinese Spoken Language Processing (ISCSLP), p. 5. IEEE (2016)
10. Ross, E.D., Monnot, M.: Neurology of affective prosody and its functional-anatomic organization in right hemisphere. Brain Lang. **104**(1), 51–74 (2008)
11. Sammler, D., Grosbras, M.H., Anwander, A., Bestelmeyer, P.E., Belin, P.: Dorsal and ventral pathways for prosody. Curr. Biol. **25**(23), 3079 (2015)
12. Jung-Beeman, M.: Bilateral brain processes for comprehending natural language. Trends Cogn. Sci. **9**(11), 512–518 (2005)
13. Kotz, S.A., Cappa, S.F., von Cramon, D.Y., Friederici, A.D.: Modulation of the lexical-semantic network by auditory semantic priming: an event-related functional MRI study. Neuroimage **17**(4), 1761 (2002)
14. Cook, R.J., Dickens, B.M., Fathalla, M.F.: World medical association declaration of Helsinki: ethical principles for medical research involving human subjects (2003)
15. Wang, H.: Modern Chinese frequency dictionary (1986)
16. Delorme, A., Makeig, S.: EEGLAB: an open source toolbox for analysis of single-trial EEG dynamics including independent component analysis. J. Neurosci. Methods **134**(1), 9–21 (2004)
17. Delorme, A., et al.: EEGLAB, SIFT, NFT, BCILAB, and ERICA: new tools for ad-vanced EEG processing. Comput. Intell. Neurosci. **1687–5265**, 10 (2011)
18. Mullen, T.R.: The dynamic brain: modeling neural dynamics and interactions from human electrophysiological recordings. Dissertations & Theses - Gradworks (2014)

19. Wernicke, C.: THE APHASIC SYMPTOM-COMPLEX: a psychological study on an anatomical basis. Arch. Neurol. **22**(3), 280–282 (1970)
20. Saur, D., et al.: Ventral and dorsal pathways for language. Proc. Natl. Acad. Sci. U. S. A. **105**(46), 18035–18040 (2008)
21. Xie, M.: The P200 component in lexical processing. Adv. Psychol. **06**(2), 114–120 (2016)
22. Pulvermüller, F., et al.: Motor cortex maps articulatory features of speech sounds. Proc. Natl. Acad. Sci. U. S. A. **103**(20), 7865–7870 (2006)
23. Möttönen, R., Dutton, R., Watkins, K.E.: Auditory-motor processing of speech sounds. Cereb. Cortex **23**(5), 1190–1197 (2013)
24. Tomasino, B., Weiss, P.H., Fink, G.R.: To move or not to move: imperatives modulate action-related verb processing in the motor system. Neuroscience **169**(1), 246–258 (2010)
25. Roy, A.C., Craighero, L., Fabbri-Destro, M., Fadiga, L.: Phonological and lexical motor facilitation during speech listening: a transcranial magnetic stimulation study. J. Physiol. Paris **102**(1–3), 101 (2008)
26. Pulvermüller, F.: Meaning and the brain: The neurosemantics of referential, interactive, and combinatorial knowledge. J. Neurolinguistics **25**(5), 423–459 (2012)
27. Barber, H.A., Kousta, S.T., Otten, L.J., Vigliocco, G.: Event-related potentials to event-related words: grammatical class and semantic attributes in the representation of knowledge. Brain Res. **1332**(1), 65–74 (2010)
28. Passeri, A., Capotosto, P., Di, M.R.: The right hemisphere contribution to semantic categorization: a TMS study. Cortex **64**, 318 (2015)

Interactions Between Modal and Amodal Semantic Areas in Spoken Word Comprehension

Bin Zhao[1], Gaoyan Zhang[1(✉)], and Jianwu Dang[1,2]

[1] Tianjin Key Laboratory of Cognitive Computing and Application,
Tianjin University, Tianjin, China
{zhaobeiyi, zhanggaoyan}@tju.edu.cn
[2] Japan Advanced Institute of Science and Technology, Ishikawa, Japan
jdang@jaist.ac

Abstract. In neurolinguistics, the controversy about whether word semantics are stored in an amodal language-specific center or distributed in modality-specific sensory-motor systems comes from two inconsistent evidences: (i) Semantic Dementia (SD) patients who got a focal brain damage in the anterior temporal lobe (ATL) exhibit a general loss of conceptual knowledge across all word categories; (ii) fMRI examinations of semantic memory found no clues in the ATL but a broad activation in the sensory and motor regions (SMR) that represent the visual and motor features of words. To settle this dispute, the current study aims to examine the whole-range brain dynamics during word processing using (i) 2-D ERP-image analysis, (ii) independent component clustering and (iii) EEG source reconstruction methods. It was found that both ATL and SMR participated in the spoken word processing by means of recurrent interaction, and the visual and motor cortex exhibited specific activation patterns for noun and verb respectively. These results suggest a hierarchical organization of word semantics that combines amodal ATL and modal SMR to form a complete concept.

Keywords: Semantics · EEG source reconstruction · Anterior temporal lobe
Sensory and motor regions

1 Introduction

In linguistic brain research, the exploration and disputation on the lexical semantic representation have always been active but it still remains as an open question. One of the most prominent disputes is whether semantics are grounded in a language-specific amodal center or embedded in modality-specific sensory and motor systems [1]. Important insights about amodal conceptual representations mainly come from semantic dementia (SD) patients who have a local damage in the left anterior temporal lobe (ATL) [2]. They suffer from a global conceptual knowledge deficit that affects the processing of all categories of objects and words [3] and all sensory-motor modalities [4]. Evidence from repetitive transcranial magnetic stimulation (rTMS) in ATL of healthy subjects also reflected a comprehensive impairment on their performance in all

© Springer Nature Switzerland AG 2018
Q. Fang et al. (Eds.): ISSP 2017, LNAI 10733, pp. 198–207, 2018.
https://doi.org/10.1007/978-3-030-00126-1_18

tasks requiring semantic knowledge [5]. These consistency regarding the role of ATL in semantic processing suggests that the ATL is critical for amodal and domain-general aspects of semantic processing [6, 7]. Contrarily, the modality-specific theory conceives that conceptual knowledge and semantics are distributed within a network of modality-specific processing regions over the sensory-motor cortex and arise from the co-activation with their sensory-motor properties [8]. As demonstrated by a number of functional magnetic resonance imaging (fMRI) and electroencephalography (EEG) studies, the sensory and motor regions (SMR) are differentially involved in response to words from different categories [9, 10]. Specifically, object nouns with visual properties tend to activate the visual occipital cortex, while action verbs with motor properties tend to activate the frontal motor and premotor cortices. This evidence conforms to the grounded cognition and embodiment theory which claims that semantics are functionally and neuroanatomically grounded in or embodied as distributed modality-specific sensory and motor systems according to the nature of the semantics. Recently, there is an emerging viewpoint that both ATL and SMS are involved in semantic processing. As the hub-and-spoke model [11] holds, ATL regions act as a transmodal conceptual hub in conceptual representation, distilling the distributed sensory-motor features of subjects and words into integrated and coherent representations.

The inconsistency of above studies might be partly explained by technical restriction. Semantic processing is a time-varying process. However, SD patient studies and low-time-resolution neuroimaging techniques are inadequate for capturing the transient variation or interaction between the modal and amodal areas if any. In view of above-mentioned facts, the current study attempts to use high-temporal-resolution EEG techniques, combining the 2-D ERP-image analysis [12], independent component decomposition and clustering algorithm [13], as well as current density reconstruction (CDR) methods [14] to reconstruct the spatiotemporal dynamics of cortical sources during semantic processing in a word-listening task, hoping for a clarification on this topic.

2 Experiment

2.1 Materials

In the listening task of this experiment, the auditory materials were 80 disyllable nouns, 80 disyllable verbs and 160 white noise segments. Here nouns refer to visual non-manipulable entities, e.g. 外衣 (coat), and verbs refer to actions, e.g. 爬山 (climb). They were all recorded at a sampling rate of 44100 Hz with 16 bits and lasted for 900 ms. Statistical analyses of the words' properties were conducted to make sure that there is no significant difference between nouns and verbs in their familiarity, concreteness, arousal and valence (ps > 0.05). Twenty right-handed Mandarin speakers (mean age = 22.3, SD = 2.1) were recruited in the listening task. The study was performed in agreement with local ethical standards. Written informed consent was obtained from all participant prior to the experiments.

2.2   Paradigm

Figure 1 shows the paradigm of this experiment. For each trial, a fixation across first appeared on a screen for 400 ms before the presentation of the auditory stimuli. After a 500-ms inter-stimulus-interval, the subjects were instructed to make a categorical judgement by pressing button 1 for words (nouns and verbs) and button 2 for noise segment that they have just heard. In the meantime, their EEG signals were recorded with 128 electrodes on the scalp at a sampling rate of 1000 Hz.

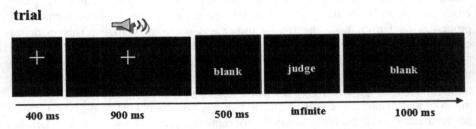

Fig. 1. Experimental paradigm

3   Methods and Results

3.1   EEG Preprocessing

The preprocessing of raw EEG data was performed using EEGLAB toolbox (http://www.sccn.ucsd.edu/eeglab) [13]. For each subject, the EEG signals were filtered at a bandwidth of 1–45 Hz and down-sampled to 256 Hz. Bad channels with abnormal fluctuation over 10% of the whole EEG range were removed before re-referencing the data to average. After rejecting EEG artifact components like eye blinks, eye movements and muscle activities with blind source separation method [15], epochs of 200 ms prior and 1000 ms after the stimulus onset were extracted and classified into noun, verb and noise groups.

3.2   ERP-Image Analysis

Traditional practice for comparing EEG results used to average all the epochs of each group and compare the 1-dimensional event-related potentials (ERPs). This study, instead, plotted 2-dimensional (2-D) ERP-images as illustrated in Fig. 2, in which the x-axis marks the time point of each trial, and the y-axis indexes the brain behavioral of all the trials in each condition. The potential value of a single trial at each time point was color-coded in accordance with the color-bar on the rightmost side. These epoch/trial lines were then sorted in the order of event response latency (reaction time) and stacked up above each other, forming a 2-D image which enables us to view the epoched data at a time-by-epoch view.

Considering the fact of noun-verb dissociation over the visual and motor cortices based on semantics [16], we examined the 2-D ERP-images for noise, noun and verb

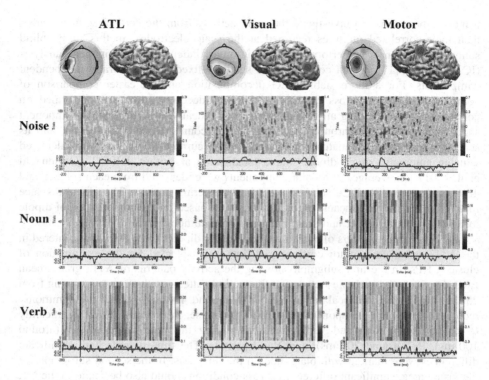

Fig. 2. ERP-image plots for noise, noun and verb conditions in the ATL, visual and motor areas. (Color figure online)

conditions from three channels that located most closely to the ATL, visual and motor cortices, as illustrated in Fig. 2. The three regions of interest were red dotted on the topography and cortical maps in the upper image. The range of y-axis is 80 trials for noun and verb and 160 trials for noise. The time range of x-axis are - 200–1000 ms. In the image, if a red vertical strap shows at a certain time point, that means a time-locked event occurs in almost all the trials and that is probably a critical event we are exploring. As we can see, in the ATL, noise-elicited activities are pretty trivial in the whole process, while nouns and verbs induced active responses from 200 ms to 600 ms with intensive red lines in this range. Some early responses around 50 ms and 100 ms could also be observed in noun and verb conditions. In the visual and motor cortices, responses to noise still exhibit no obvious clues for any time-locked event. In comparison, nouns evoked larger positive responses during 80–100 ms, 220–250 ms, 300–360 ms and 410–430 ms in the visual cortex, and verbs activated large potentials in the motor cortex during 230–300 ms, and even stronger activities in a wide range of 400–500 ms.

3.3 Independent Component Clustering

In that the scalp-recorded data is a mixture of activities from multiple underlying cortical sources and non-brain sources with different locations, extent and orientations, it makes more sense to investigate the brain activity from the cortical sources, rather than their correlated mixtures recorded at the scalp electrodes. In this study, blind source separation was performed using infomax independent component analysis (ICA) method [17], which could provide spatially fixed and temporally independent components. For a more stable EEG decomposition and an easier comparison of component behavior across conditions, the ICA decomposition was performed on continuous data including all stimulus categories. Ideally, an independent component should account for synchronous activity within a connected cortical domain, and its scalp projection should match a single equivalent current dipole that was calculated with an automatic single dipole source localization algorithm, DIPFIT (Oostenveld et al. 2001). On average, 24 dipolar components were identified from each subject, and 480 components for all 20 subjects went into clustering process. The global distance matrix for the use of 'kmeans' algorithm was built based on the combination of dipole locations, spectral power and scalp maps of those components.

In Fig. 3, three clusters of equivalent component dipoles (blue dots) that centered in the ATL, visual and motor cortices (red dot) were selected for the examination of cluster ERPs. In the first subgraph showing the activity patterns of ATL, grand mean ERPs for noun and verb present similar fluctuation patterns with a hump spanning from 250 ms to 550 ms and peaking in the N400 (around 400 ms) period, the commonly recognized semantic component. The ERPs for noise, in contrast, differ significantly ($p < 0.05$) with noun and verb conditions during the time range of N1 (around 100 ms), P2 (around 200 ms), and N400 periods. The time periods with significant differences were marked with black line on the time axis. In the second sub-graph for the visual area, significant differences across-conditions could also be found in the N1, P2 and periods around 500 ms and after 600 ms. It can be noticed that ERP waves diverge earlier than 50 ms and even before the word onset (0 ms). This might be explained by the visual responses to the fixation across that occur 400 ms before the stimulus onset. Specifically, in the N400 semantic processing period, noun (with visual object meaning) elicited positive fluctuation while noise and verb elicited negative fluctuation. This difference, though not significant, might be interpreted as visual-associated response specific for object-related nouns rather than meaningless noise or action verbs. In the third sub-graph for motor areas, positive peaks around 100 ms were found for the noise and noun conditions, but not for the verb condition. Instead, the activity for verbs presented significant ($p < 0.05$) negative peaks around 320 ms and 770 ms than its noun and noise counterparts.

3.4   Current Density Reconstruction

To trace the information flow among the interest cortical sources, current density reconstruction was performed with the constraint of standardized Low Resolution Electromagnetic Tomography, which is capable of presenting the activated sources that distributed on a cortical map and providing the activation extent at a certain time point. This method provides a direct and intuitive view of brain activities on the cortex along the whole-time range of semantic process.

To further investigate the semantic relations with ATL and the visual and motor system, In Fig. 4, the CDR results of noun and verb processing are displayed during the

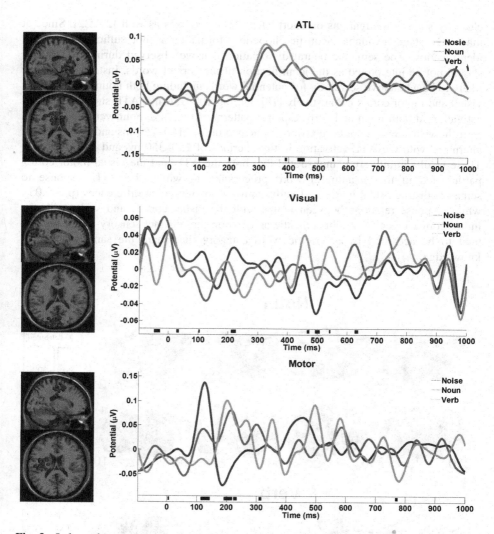

Fig. 3. Independent component clusters and grand mean component ERPs for noise, noun and verb conditions in the ATL, visual and motor areas. For each sub-graph, the left MRI (Montreal Neurological Institute) images (sagittal and top view) show the clustered components (blue dots) that centered in the ATL, visual and motor areas (red dots), and the right plot shows the grand mean component cluster ERPs for each condition with significant group differences marked black on the time axis. (Color figure online)

period of 0–1000 ms, respectively. As illustrated, after stimulus onset, ATL was activated, together with auditory cortex in the superior temporal area in both noun and verb conditions. Then in the 50–100 ms range, current in response to nouns flowed to the visual cortex while it flowed to the motor cortex in response to verbs. Interestingly, early ERP components around 50 ms and 100 ms have been speculated as semantic clues by some investigations on short-latency time courses as well [21, 22]. Since at that early stage, available acoustic-phonetic information is not sufficient for word identification. The semantic neural dissociation that we observed during 50–100 ms should not be interpreted as the identification of the correct words. Instead, it is more likely to be a tentative searching for potential words including both nouns and verbs in visual and motor cortex exhaustively [18]. During 100–150 ms, ATL showed repeated activation in both noun and verb response patterns. Afterward, noun-verb dissociation over the visual and motor cortex were replicated during 150–250 ms and 350–450 ms, alternated with the ATL activation in time periods of 250–350 ms and 450–1000 ms. These results intuitively demonstrated that both the ATL and the visual-motor cortex participated in the meaningful word processing, in which the ATL response to semantic-distinguished nouns and verbs showed no significant differences ($p > 0.05$), while response relations between nouns with the visual cortex, and verbs with the motor cortex revealed a tight semantic association that distinct sensory-motor regions tend to be activated by a specific word category that share the same conceptual knowledge.

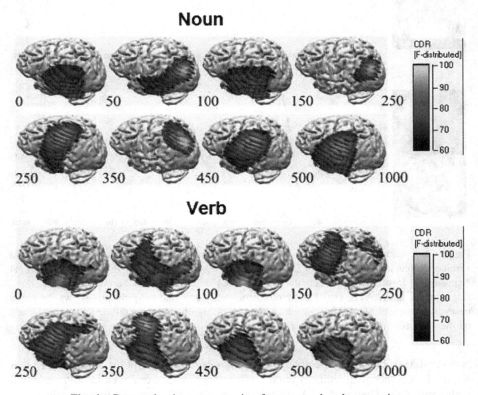

Fig. 4. Current density reconstruction for noun and verb processing.

4 Discussion and Conclusion

This study aims to address the issue regarding whether semantic representation is amodal or modality-specific entities. Amodal theory assumes that sensory and motor information from the environment has been transformed into an amodal symbolic representation format, which is probably located in the ATL and separated from sensory and motor systems. In contrast, modality-specific theory believes that conceptual representation are functionally and neuroanatomically grounded in sensory and motor representations. While in our studies, the results of 2-D ERP-images, the equivalent dipoles of independent component clustering and the current density flow among cortical sources all indicate that regions within the anterior temporal lobes and sensory motor regions are both critical nodes in the neural network for representing conceptual knowledge and semantic processing.

From plenty of evidence in previous research [9, 10], the role of sensory and motor regions in the representation of specific modality features is widely accepted. What is relatively uncertain is the ATL part in semantic processing. Some authors supporting amodal theory [6] suggested that language processing and conceptual representation are functionally separated from perceptual and action systems of the brain. The amodal symbolic system would interface with sensorimotor systems only for receiving its input or passing on its output, but otherwise maintain functional separation from action and perception systems. To elucidate the functional connectivity of ATL and the functional network underlying semantic cognition, Jackson et al. [19] performed a psychophysiological interactions (PPI) analysis to compare the activity between resting-state and active semantic tasks. The PPI analysis suggested a core semantic network where the ventral ATL and anterior middle temporal gyrus were shown to connect to areas responsible for multimodal semantic cognition, and the anterior superior temporal gyrus was connected to a distinct set of auditory and language-related areas. The study also showed additional connectivity of the ATL to regions of occipital and frontal cortex in the semantic tasks than the resting state, which is in consistent with our results, supporting that both ATL and sensory-motor areas, including occipital visual cortex and frontal-motor cortex are sensitive to semantic processing. In a study by Paul [11], semantic dementia patients were asked to assign abstract visual stimuli to two categories that conformed to a family resemblance structure in which no individual stimulus features were fully diagnostic, thus the task required participants to form representations that integrate multiple features into a single concept. However, SD patients failed to integrate information but only respond on the basis of individual features. This study reveals that ATL undertakes the critical computational function of integrating disparate sources of information into novel coherent concepts. The crucial role of ATL in transmodal semantic representation also fit with a recent tractography study demonstrating the convergence of multiple white matter pathways into the ATL where the structural connectivity is ideal for blending different sources of information into integrated, coherent concepts [20]. These evidences have been developed into an alternative framework for conceptual knowledge termed the "hub-and spoke" model [2, 21], which holds that in addition to the modality-specific information sources ("spokes"), a modality-invariant conceptual representation "hub" is also required to capture deeper conceptual patterns across all sensory-motor and verbal modalities. This hub is probably located in the ATL.

By a bold conjecture [22], the modality-specific brain regions may be associated with specific object or action properties, and the ATL is more likely to carry a higher-order information that abstract away from concrete semantic attributes. In this view, word semantics are hierarchically organized from distributed low-level representations to a generalized high-level integration. These speculations still require careful verification in future investigations.

Acknowledgements. The research is supported partially by the National Basic Research Program of China (No. 2013CB329301), and the National Natural Science Foundation of China (No. 61233009 and 61503278). The study is supported partially by JSPS KAKENHI Grant (16K00297).

References

1. Kiefer, M., Pulvermüller, F.: Conceptual representations in mind and brain: theoretical developments, current evidence and future directions. Cortex **48**, 805 (2012)
2. Patterson, K., Nestor, P.J., Rogers, T.T.: Where do you know what you know? The representation of semantic knowledge in the human brain. Nat. Rev. Neurosci. **8**, 976–987 (2007)
3. Hoffman, P., Ralph, M.A.L.: Reverse concreteness effects are not a typical feature of semantic dementia: evidence for the hub-and-spoke model of conceptual representation. Cereb. Cortex **21**, 2103 (2011)
4. Bozeat, S., Lambon Ralph, M.A., Patterson, K., Garrard, P., Hodges, J.R.: Non-verbal semantic impairment in semantic dementia. Neuropsychologia **38**, 1207–1215 (2000)
5. Pobric, G., Jefferies, E., Ralph, M.A.L.: Anterior temporal lobes mediate semantic representation: mimicking semantic dementia by using rTMS in normal participants. Proc. Natl. Acad. Sci. U S A **104**, 20137–20141 (2007)
6. Rogers, T.T., Hocking, J., Noppeney, U., Mechelli, A., Gornotempini, M.L., Patterson, K., et al.: Anterior temporal cortex and semantic memory: reconciling findings from neuropsychology and functional imaging. Cogn. Affect. Behav. Neurosci. **6**, 201 (2006)
7. Review, P.: Structure and deterioration of semantic memory: a neuropsychological and computational investigation. Psychol. Rev. **111**, 205–235 (2004)
8. Martin, A.: The representation of object concepts in the Brain. Annu. Rev. Psychol. **58**, 25 (2007)
9. Moseley, R.L., Pulvermüller, F.: Nouns, verbs, objects, actions, and abstractions: local fMRI activity indexes semantics, not lexical categories. Brain Lang. **132**, 28 (2014)
10. Zhao, B., Dang, J., Zhang, G.: EEG source reconstruction evidence for the noun-verb neural dissociation along semantic dimensions. Neuroscience **359**, 183–195 (2017)
11. Paul, H., Evans, G.A.L., Lambon, R.M.A.: The anterior temporal lobes are critically involved in acquiring new conceptual knowledge: evidence for impaired feature integration in semantic dementia✩. Cortex **50**, 19 (2014)
12. Makeig, S., Westerfield, M., Enghoff, S., Townsend, J., Courchesne, E., Sejnowski, T.J.: Dynamic brain sources of visual evoked responses. Science **295**, 690–694 (2002)
13. Delorme, A., Makeig, S.: EEGLAB: an open source toolbox for analysis of single-trial EEG dynamics including independent component analysis. J. Neurosci. Methods **134**, 9 (2004)

14. Pascual-Marqui, R.D.: Standardized low-resolution brain electromagnetic tomography (sLORETA): technical details. Methods Find. Exp. Clin. Pharmacol. **24**(Suppl D), 5–12 (2002)

15. Makeig, S., Jung, T.P., Bell, A.J., Ghahremani, D., Sejnowski, T.J.: Blind separation of auditory event-related brain responses into independent components. Proc. Natl. Acad. Sci. U. S. A. **94**, 10979–10984 (1997)

16. Zhao, B., Dang, J., Zhang, G.: Investigation of noun-verb dissociation based on EEG source reconstruction. In: Signal and Information Processing Association Summit and Conference, pp. 1–7 (2017)

17. Bell, A.J., Sejnowski, T.J.: An information-maximization approach to blind separation and blind deconvolution. Neural Comput. **7**, 1129–1159 (1995)

18. Zhao, B., Dang, J., Zhang, G.: EEG evidence for a three-phase recurrent process during spoken word processing. Presented at the 10th ISCSLP, Tianjin, China (2016)

19. Jackson, R.L., Hoffman, P., Pobric, G., Ralph, M.A.L.: The semantic network at work and rest: differential connectivity of anterior temporal lobe subregions. J. Neurosci. Official J. Soc. Neurosci. **36**, 1490 (2016)

20. Binney, R.J., Parker, G.J., Lambon Ralph, M.A.: Convergent connectivity and graded specialization in the rostral human temporal lobe as revealed by diffusion-weighted imaging probabilistic tractography. J. Cogn. Neurosci. **24**, 1998–2014 (2012)

21. Lambon Ralph, M.A., Sage, K., Jones, R.W., Mayberry, E.J.: Coherent concepts are computed in the anterior temporal lobes. Proc. Natl. Acad. Sci. U.S.A. **107**, 2717–2722 (2010)

22. Huth, A.G., De, H.W.A., Griffiths, T.L., Theunissen, F.E., Gallant, J.L.: Natural speech reveals the semantic maps that tile human cerebral cortex. Nature **532**, 453–458 (2016)

Speech Disorder

Acoustic Analysis of Mandarin Speech in Parkinson's Disease with the Effects of Levodopa

Wentao Gu[1(✉)], Ping Fan[1], and Weiguo Liu[2]

[1] Nanjing Normal University, Nanjing 210097, China
wtgu@njnu.edu.cn, fpshida2010@126.com
[2] Nanjing Brain Hospital Affiliated to Nanjing Medical University,
Nanjing 210029, China
liuweiguol111@sina.com

Abstract. This study investigated prosodic and articulatory characteristics of parkinsonian speech by acoustic analysis of the read speech from 21 Mandarin-speaking Parkinson's Disease (PD) patients before and after administration of levodopa medication, and 21 age- and gender-matched healthy controls (HC). PD exhibited reduced F0 variability and increased minimum intensity during the closures of stops than HC. For females, PD also showed smaller vowel space area and vowel articulation index than HC. Administration of levodopa increased the mean, the max, and the range of F0, bringing them closer to those of HC, but little effect was found on other acoustic parameters. Correlation analysis between acoustic parameters and physiological/pathological indices of PD showed that the only significant positive correlation was between pause ratio and UPDRS III score. The findings on acoustic differences between PD and HC can potentially be applied to diagnosis and speech therapy for PD.

Keywords: Parkinson's disease · Mandarin · Levodopa · Acoustic analysis Prosody · Articulation

1 Introduction

Parkinson's disease (PD) is one of the neurodegenerative diseases in the middle-aged and elderly people. It is characterized by progressive loss of dopaminergic neurons in the central nervous system, especially substantia nigra pars compacta in basal ganglia. Typical symptoms of motor impairments in PD patients include bradykinesia, hypokinesia, akinesia, muscle rigidity and rest tremor (Brodal 1998; Lang and Lozano 1998).

In addition, 70%–90% of PD patients also suffer from hypokinetic dysarthria which may have developed for years before the appearance of obvious clinical motor symptoms and hence may be an indicator for early diagnose of PD (Ho et al. 1999; Harel et al. 2004; Raming et al. 2008; Miller et al. 2011; Rusz et al. 2011).

Up to now, levodopa therapy has been the most widely adopted pharmacological treatment which greatly alleviates limb motor symptoms of PD. However, its effects on hypokinetic dysarthria in speech production are still controversial.

© Springer Nature Switzerland AG 2018
Q. Fang et al. (Eds.): ISSP 2017, LNAI 10733, pp. 211–224, 2018.
https://doi.org/10.1007/978-3-030-00126-1_19

1.1   Hypokinetic Dysarthria of PD

Hypokinetic dysarthria is common in extrapyramidal system diseases like PD. The hypokinetic dysarthria in PD is manifested in all dimensions of speech production including respiration, phonation, and articulation, and is perceptually characterized by monotone, monoloudness, abnormal speech rate, hoarse voice, imprecise articulation, and so on (Goberman and Coelho 2002).

Prosodic features of speech mainly include pitch, length and loudness, which acoustically correspond roughly to fundamental frequency (F0), duration, and intensity of speech signals. Previous studies noted that PD's speech had higher F0, narrower F0 range (Canter 1963; Ma and Hoffmann 2010; Rigaldie et al. 2006) and smaller F0 variation than those of healthy controls (HC) (Gamboa et al. 1997; Jiménez-Jiménez et al. 1997; Harel et al. 2004; Viallet et al. 2008; Rusz et al. 2011; Bowen et al. 2013; Galaz et al. 2016).

However, there are diverging results on speech rate. In previous studies, PD patients exhibited faster rate (Weismer 1984; Flint et al. 1992; Hammen et al. 1996) or slower rate (Ludlow et al. 1987; Chenausky et al. 2011) than HC, or no significant difference in speech rate was found between PD and HC due to great individual differences among PD patients (Ma and Hoffmann 2010; Skodda and Schlegel 2010; Chenausky et al. 2011). Most previous studies reported monoloudness in PD's speech, with a few exceptions showing no difference in loudness between PD and HC (Canter 1963; Ma et al. 2010b).

PD patients also have impairments in articulation of vowels and consonants. Logemann and Fisher (1981) argued that articulation deficits (incomplete closure) of stops /b, p, d, t, k/ might result from weakened motions of lips, tongue and palate. This was verified by such indices as intensity during closure (IDC), intensity of closure relative to that of the following vowel, and intensity of closure relative to the peak amplitude of the syllable (Ackermann and Ziegler 1991; Chenausky et al. 2011; Dromey and Bjarnason 2011; Karlsson et al. 2014).

Impairments in vowel articulation are usually characterized by the indices calculated from F1 and F2 formants of the corner vowels. A frequently-used index is Vowel Space Area (VSA), which is reduced in PD and hence may be used as an indicator of disease progression of PD (Turner et al. 1995; Weismer et al. 2001; Skodda et al. 2012). Likewise, another index Vowel Articulation Index (VAI) is smaller in PD than in HC (Roy et al. 2009; Sapir et al. 2010; Skodda et al. 2011b). Both indices suggest reduced amplitude and velocity of the motions of the articulators such as tongue and lips in PD patients.

1.2   Levodopa Treatment to PD

Levodopa is by far the most widely adopted and the most effective therapy to early PD. Although significant improvement on limb motor disorders after levodopa treatment has been well documented, its effects on speech disorders are still controversial.

Previous perceptual studies of PD's speech reported significant improvements in prosody, voice quality, articulation, and hence the overall intelligibility of speech, after levodopa treatment (Mawdsley et al. 1971; Nakano et al. 1973; Wolfe et al. 1975).

However, acoustic studies showed diverging results on the effects of levodopa on PD's speech. While some studies found positive effects in increasing F0 variability (Harel et al. 2004; Bowen et al. 2013) and F0 mean (Rigaldie et al. 2006; Viallet et al. 2008) as well as improving other prosodic features (De Letter et al. 2003, 2005), other studies did not find any significant improvement in speech production (Goberman, Coelho, and Robb 2002, 2005; De Letter et al. 2006; Ho et al. 2008; Skodda et al. 2010b; Skodda et al. 2011a).

The failure to improve PD's speech by levodopa treatment seems to suggest that speech production is controlled not only by dopaminergic system but also by non-dopaminergic system (Bejjani et al. 2000; Baumgartner et al. 2001; Goberman, Coelho, and Robb 2005; Skodda et al. 2010b; Skodda et al. 2011a).

1.3 Purpose of the Study

The above-mentioned literatures all dealt with the PD patients speaking western languages, and there were also some studies dealing with Cantonese-speaking PD patients (Ma 2009; Ma et al. 2010a; Ma et al. 2010b; Ma et al. 2011). However, few research on Mandarin-speaking PD patients has been conducted.

To fill in the gaps, the present study aimed to investigate the acoustic characteristics of speech in Mandarin-speaking PD patients, to explore the possibility of early diagnosis of PD. Meanwhile, the effects of pharmacological levodopa treatment on PD's speech were tested by comparing PD's speech produced at the pre-medication (OFF) and post-medication (ON) states. By comparison with previous studies, we will examine whether there are language-dependent characteristics in PD's speech.

2 Method

Two groups of participants were recruited in the experiment, i.e., 21 patients with idiopathic PD, and 21 age- and gender-matched healthy controls. There were 12 males and 9 females in each group.

For a controlled experiment, it is critical to have a homogeneous group of PD patients. Neurologists have made a series of assessment scales for PD, such as the Unified Parkinson's Disease Rating Scale (UPDRS) and the Mini-Mental State Examination (MMSE). The 3rd part of UPDRS gives the UPDRS III score representing the severity of motor disability, while the 5th part of UPDRS gives the UPDRS V score (i.e., the modified H&Y scale) representing the overall severity of the disease. MMSE is an examination to measure cognitive impairments. In addition to age, the modified H&Y scale and MMSE were used to select the homogeneous PD patients.

In the present study, all PD participants were at the mild to moderate stage of PD (i.e., modified H&Y scale between 1 and 3) without dementia (MMSE > 24), and had no experience of speech therapy. All HC participants were physically and mentally healthy without any speech disorder. The means and standard deviations of the physiological/pathological indices of the PD and HC participants are listed in Table 1. For each patient, UPDRS III was scored before and after the administration of levodopa (named the OFF and ON states, respectively). As shown, the mean UPDRS III score

Table 1. Statistics of the physiological/pathological indices of the participants (mean ± SD.).

	PD	HC
Age (year)	62.95 ± 6.65	62.71 ± 5.53
Modified H&Y	2.17 ± 0.75	N/A
MMSE	27.71 ± 2.25	N/A
UPDRS III (OFF)	29.33 ± 14.54	N/A
UPDRS III (ON)	18.57 ± 12.13	N/A
Disease duration (year)	5.13 ± 3.10	N/A

decreased after the administration of levodopa, suggesting an improvement of motor abilities.

The material used in this study was a short story *The North Wind and the Sun* composed of 185 syllables. Speech recording was conducted in a quiet room after the participants got familiar with the material. Each participant read aloud the story at his/her normal speaking rate and comfortable pitch and loudness. The speeches were recorded using the portable digital recorder Zoom H4n, digitized at 44.1 kHz with 16-bit precision.

For each PD participant, the speech at the OFF state was firstly collected at least 12 h after the last administration of medication to minimize the effects of levodopa. Then, after receiving the assessment scales including UPDRS and MMSE, the PD participant took medication with 1.5 times of his/her usual dose. About 60–90 min later, when the PD participant felt in his/her best condition, the PD's speech at the ON state was collected.

3   Results

All speech data were analyzed using the Praat software. For each acoustic parameter, we conducted t-tests with Bonferroni correction to compare between the HC group and the two states (OFF and ON) of the PD group.

3.1   Prosodic Characteristics

Fundamental Frequency. Five F0 parameters defined on the whole discourse, i.e., F0 mean, F0 min, F0 max, F0 range, and F0 standard deviation (F0std), were measured in semitone (St) with reference to 50 Hz.

For each parameter, paired t-tests were conducted between the ON and OFF states of PD, whereas independent t-tests were conducted between HC and each state of PD. Figure 1 shows the statistical results of all five F0 parameters. On the one hand, both ON and OFF states of PD show significantly lower F0std than HC, coinciding with the 'monotone' subjective impression of PD's speech. On the other hand, while there is no significant difference in F0 min between the two states of PD, the mean, the max, and the range of F0 are significantly higher in the ON state than in the OFF state, suggesting an increase in high pitch after the administration of levodopa.

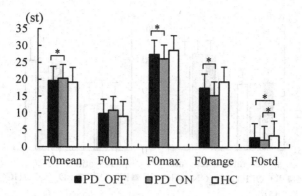

Fig. 1. Comparison of F0 parameters. (In the figures hereinafter: $*p < 0.05$, $**p < 0.01$, $***p < 0.001$; Error bars represent standard errors.)

Temporal Parameters. Temporal parameters measured in this study include speech rate, articulation rate (i.e., the rate of the articulated portions of speech), pause ratio, mean duration of vocalic intervals (V_dur), mean duration of consonantal intervals (C_dur), percentage of vocalic durations in the total utterance (%V), standard deviation of vocalic durations (ΔV), standard deviation of consonantal durations (ΔC), and voice onset time (VOT) of aspirated stop /th/. Here, a vocalic interval is the section between the onset and the offset of a series of connected vowels/glides, while a consonantal interval is the section between the onset and the offset of a series of connected consonants.

As shown in Fig. 2, there is no significant difference in any of the nine parameters between groups or states, indicating that these temporal parameters are not affected by the PD symptoms.

3.2 Articulatory Characteristics

Stops. The minimum intensities during the closures of Mandarin stops "b, p, d, t, g" (in IPA, /p, p^h, t, t^h, k/, correspondingly) were calculated. As shown in Fig. 3, for each of the five stops, PD is significantly higher than HC, while there is no significant difference between the ON and OFF states of PD.

Vowels. For three corner vowels /a/, /i/ and /u/, two formants F1 and F2 were measured from the central 60% interval (also the most stable part) of the vowels. The triangular Vowel Space Area (tVSA) (Liu et al. 2005) and Vowel Articulation Index (VAI) (Roy et al. 2009; Sapir et al. 2010) were calculated as follows.

$$tVSA = ABS\{[F1[i] \times (F2[a] - F2[u]) \\ + F1[a] \times (F2[u] - F2[i]) \\ + F1[u] \times (F2[i] - F2[a])]/2\} \tag{1}$$

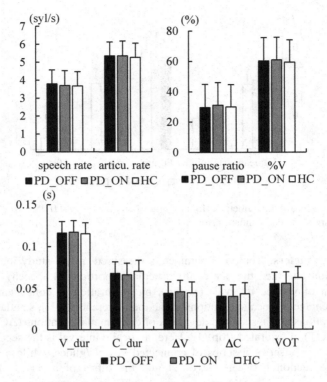

Fig. 2. Comparison of temporal parameters.

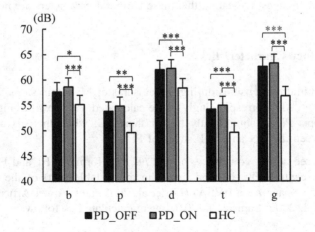

Fig. 3. Comparison of the minimum intensities during the closures of stops.

$$VAI = (F2[i] + F1[a])/(F1[i] + F1[u] + F2[u] + F2[a]) \qquad (2)$$

Both indices represent the degree of dispersion for the formants of vowels. Smaller tVSA and VAI indicate more centralized vowels (by definition the minimum of VAI is 0.5).

Because formants are determined by the length of the resonance cavity which varies with gender, we analyzed male and female participants separately. Figure 4 shows the formant distribution of the three corner vowels. The triangles constituted by solid, dashed, and dotted lines indicate the mean tVSAs for HC, the ON state of PD, and the OFF state of PD, respectively. Figure 5 further shows the statistical results of tVSA and VAI. For both indices, significant decreases in PD (both ON and OFF states) relative to HC are observed only for female participants.

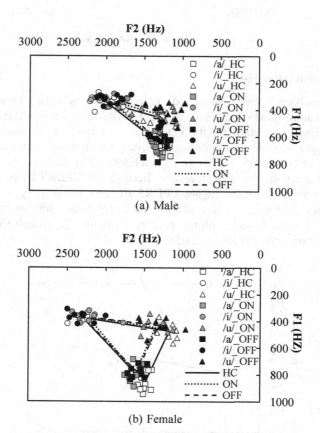

(a) Male

(b) Female

Fig. 4. Formant distribution of corner vowels /a/, /i/, and /u/.

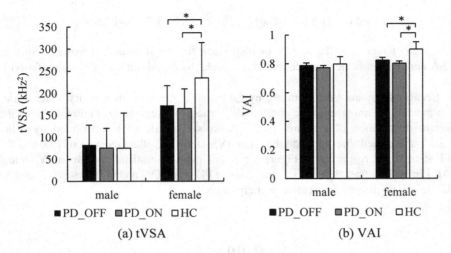

Fig. 5. Comparison of tVSA and VAI.

3.3   Correlation Analysis

Correlation analyses were conducted between all acoustic parameters and physiological/pathological indices such as age, UPDRS III score, modified H&Y scale, MMSE, and disease duration for the PD participants at the OFF state. The results showed that there was only a significant positive correlation between pause ratio and UPDRS III score (r = 0.505, p < 0.05), as exhibited in Fig. 6.

This positive correlation is interpretable. Because the UPDRS III score represents the severity of motor disorders, higher UPDRS III score is generally associated with more difficulties in moving the articulators and lower breath support caused by the rigidity of chest wall, hence resulting in more disfluency of speech (Solomon and Hixon 1993; Goberman and Elmer 2005).

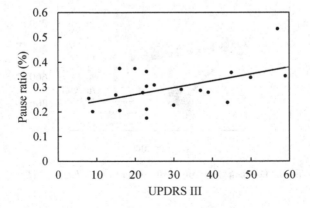

Fig. 6. Correlation between UPDRS III score and pause ratio.

4 Discussion

4.1 Acoustic Characteristics of PD's Speech

The results of acoustic analyses have shown smaller F0 variation, higher intensity during the closures of stops, and smaller tVSA as well as VAI in PD than in HC.

The finding on reduced F0 variability coincides with the 'monotone' subjective impression as well as the results of previous acoustic studies (Gamboa et al. 1997; Jiménez-Jiménez et al. 1997; Goberman and Coelho 2002; Harel et al. 2004; Galaz et al. 2016). The reduced F0 variability may be a result of decreased range of motion of vocal folds (Goberman and Coelho 2002), and is possibly the most effective and objective index for distinguishing PD's speech from HC's (Rusz et al. 2011), though Metter and Hanson's (1986) finding that F0 variability decreases further with the progression of PD is neither supported by our results nor by those previous studies showing little correlation between F0 variability and disease duration (Gamboa et al. 1997; Holmes et al. 2000; Skodda et al. 2011a).

For temporal parameters, there was no significant difference between PD and HC. This coincides with Metter and Hanson (1986) and Blanchet and Snyder (2009) in that PD patients can speak at a too fast or too slow rate in comparison with HC.

The minimum intensity during the closures of stops "b, p, d, t, g" (in IPA, /p, p^h, t, t^h, k/) was significantly higher in PD than in HC. This was manifested as spirantization of stops, because airflow escaped from the incomplete obstruction of the articulators as a result of muscle impairments in PD.

For females, tVSA and VAI were significantly larger in PD than in HC. This was manifested as centralization of vowels as a result of limited amplitude and range of motion of the articulators due to muscle impairments in PD. For males, however, no significant difference was found. The reason why this deficit was observed only in females might be that among healthy speakers the articulation of females is usually more standard than that of males for social reasons (Munson and Babel, in press); hence, muscle impairments might have more serious impacts on the females' tongue positioning and thus vowel quality than on males'.

4.2 Effects of Levodopa on PD's Speech

Acoustic comparison of PD's speech before and after medication showed slight but significant increases in the mean, the max, and the range of F0. The increase of the mean F0 coincides with the reports in Rigaldie et al. (2006) and Viallet et al. (2008).

Bowen et al. (2013) noted that PD progression was caused by the loss of dopaminergic neuron in basal ganglia. As a typical motor disorder, hypokinetic dysarthria is affected by dopamine deficits, and supplements of dopamine can in principle alleviate hypokinetic dysarthria. It is to be noted that the observed positive effect on F0 may be related to the increased dose of medicine in this study – the PD participants were asked to take medication 1.5 times of their usual dose (hence at least 200 mg).

However, besides F0 levels, no other acoustic parameters of speech improved after the medication. In other words, the positive effects of the administration of levodopa on speech were very limited. This coincides with the previous finding that the axial symptoms including speech, gait, and postural stability could not be alleviated efficiently by levodopa (Klawans 1986; Bonnet et al. 1987).

In fact, many previous studies have proposed that both dopaminergic and non-dopaminergic systems should be responsible for hypokinetic dysarthria of PD. For example, Braak et al. (2004) argued that the brainstem motor system including vagus and glossopharyngeal nerves were likely to be affected in the earlier stages of PD (i.e., speech disorders existed preclinically), whereas dopaminergic deficiency occurred only in the later stages. During recent years, many factors other than dopamine deficiency that possibly account for speech disorders in PD have been studied, such as underscaling of vocal efforts (Solomon and Robin 2005; Lafargue et al. 2008), deficits in internal cueing (Goberman and Elmer 2005; Möbes et al. 2008), sensory processing (Liu et al. 2012) and so on, as reviewed in Sapir (2014).

4.3 Crosslinguistic Comparison of PD's Speech

Our acoustic analyses of Mandarin speech in PD patients showed reduced F0 variability, and increased minimum intensity during the closures of stops in comparison with HC. For females, PD also showed smaller tVSA and VAI than HC. These basically coincided with the previous findings on PD's speech in other languages. Especially, the reduced F0 variability was also observed in PD patients speaking Cantonese, another Chinese tone language (Ma 2009; Ma et al. 2010a; Ma et al. 2010b). For temporal parameters, the present study did not find any significant difference between PD and HC. This actually also coincides with the fact that the results on temporal parameters diverged greatly in previous studies on other languages due to various differences in individual PD patients, types of speech material, and methods of measurement. Therefore, no language-specific deficit in acoustic manifestations of speech has been obviously observed in PD patients.

5   Conclusions

The present study examined acoustic characteristics of PD's Mandarin speech by comparing between 21 PD patients before and after administration of levodopa medication, and 21 age- and gender-matched healthy controls. PD's speech exhibited smaller F0 variability and higher minimum intensity during the closures of stops. For females, PD also showed smaller tVSA and VAI than HC (but for males, there was little difference). Correlation analysis between acoustic parameters and physiological/pathological indices of PD showed that the only significant positive correlation was between pause ratio and the UPDRS III score. Administration of levodopa increased the mean, the max, and the range of F0 slightly but significantly, bringing them closer to HC's values, but no positive effect was observed on articulatory parameters and other prosodic parameters.

Since speech production is a complicated process involving precise coordination of a set of articulators under the control of various muscles, hypokinetic dysarthria of PD is manifested in both prosodic and articulatory features of speech. The results of this study suggest that F0 variability, minimum intensity during the closures of stops, and tVSA as well as VAI are potentially effective acoustic indices for diagnosis of PD in Mandarin speakers, and are expected to be also useful for speech therapy of PD. The effects of administration of levodopa on speech seem to be limited only in pitch, and more research is needed to further explore the mechanisms.

Acknowledgements. This work is supported by the Major Program of the National Social Science Fund of China (13&ZD189), National Natural Science Foundation of China (81571348), Jiangsu Natural Science Foundation (SBK2015022028), Jiangsu Provincial Key Research and Development Program (BE2016614) and the project for Jiangsu Higher Institutions' Excellent Innovative Team for Philosophy and Social Sciences (2017STD006). We also thank the Neurology Department of the Second Affiliated Hospital of Soochow University for their helping collect a part of PD's speech data.

References

Ackermann, H., Ziegler, W.: Articulatory deficits in parkinsonian dysarthria: an acoustic analysis. J. Neurol. Neurosurg. Psychiatry **54**(12), 1093–1098 (1991)

Baumgartner, C.A., Sapir, S., Ramig, T.O.: Voice quality changes following phonatory-respiratory effort treatment (LSVT) versus respiratory effort treatment for individuals with Parkinson disease. J. Voice **15**(1), 105–114 (2001)

Bejjani, B.P., Gervais, D., Arnulf, I., et al.: Axial parkinsonian symptoms can be improved: the role of levodopa and bilateral subthalamic stimulation. J. Neurol. Neurosurg. Psychiatry **68**(5), 595–600 (2000)

Blanchet, P.G., Snyder, G.J.: Speech rate deficits in individuals with Parkinson's disease: a review of the literature. J. Med. Speech-Lang. Pathol. **17**(1), 1–7 (2009)

Bonnet, A.M., Loria, Y., Saint-Hilaire, M.H., et al.: Does long-term aggravation of Parkinson's disease result from nondopaminergic lesions? Neurology **37**(9), 1539–1542 (1987)

Bowen, L.K., Hands, G., Pradhan, L.S., Stepp, C.E.: Effects of Parkinson's disease on fundamental frequency variability in running speech. J. Med. Speech-Lang. Pathol. **21**(3), 235–244 (2013)

Braak, H., Ghebremedhin, E., Rüb, U., et al.: Stages in the development of Parkinson's disease-related pathology. Cell Tissue Res. **318**(1), 121–134 (2004)

Brodal, P.: The Central Nervous System: Structure and Function, 2nd edn. Oxford University Press, New York (1998)

Canter, G.J.: Speech characteristics of patients with Parkinson's disease: I. Intensity, pitch, and duration. J. Speech Hear. Disord. **28**(3), 221–229 (1963)

Chenausky, K., Macauslan, J., Goldhor, R.: Acoustic analysis of PD speech. Parkinson's Disease, 435232 (2011)

De Letter, M., Santens, J.V., Borsel, P.: The effects of levodopa on word intelligibility in Parkinson's disease. J. Commun. Disord. **38**(3), 187–196 (2005)

De Letter, M., Santens, P., De, B.M., et al.: Levodopa-induced alterations in speech rate in advanced Parkinson's disease. Acta Neurol. Belg. **106**(1), 19–22 (2006)

De Letter, M., Santens, P., Van, B.J.: The effects of levodopa on tongue strength and endurance in patients with Parkinson's disease. Acta Neurol. Belg. **103**(1), 35–38 (2003)

Dromey, C., Bjarnason, S.: A preliminary report on disordered speech with deep brain stimulation in individuals with Parkinson's disease. Parkinson's Disease, 796205 (2011)

Flint, A.J., Black, S.E., Campbell-Taylor, I., et al.: Acoustic analysis in the differentiation of Parkinson's disease and major depression. J. Psycholinguist. Res. **21**(5), 383–399 (1992)

Galaz, Z., Mekyska, J., Mzourek, Z., et al.: Prosodic analysis of neutral, stress-modified and rhymed speech in patients with Parkinson's disease. Comput. Methods Programs Biomed. **127**, 301–317 (2016)

Gamboa, J., Jimenez-Jimenez, F.J., Nieto, A., et al.: Acoustic voice analysis in patients with Parkinson's disease treated with dopaminergic drugs. J. Voice **11**(3), 314–320 (1997)

Goberman, A., Coelho, C.: Acoustic analysis of parkinsonian speech I: speech characteristics and L-Dopa therapy. Neurorehabilitation **17**(3), 237–246 (2002)

Goberman, A., Coelho, C., Robb, M.: Phonatory characteristics of Parkinsonian speech before and after morning medication: the ON and OFF states. J. Commun. Disord. **35**(3), 217–239 (2002)

Goberman, A., Coelho, C., Robb, M.: Prosodic characteristics of Parkinsonian speech: the effect of levadopa-based medication. J. Med. Speech-Lang. Pathol. **13**(1), 51–68 (2005)

Goberman, A., Elmer, L.W.: Acoustic analysis of clear versus conversational speech in individuals with Parkinson disease. J. Commun. Disord. **38**(3), 215–230 (2005)

Hammen, V.L., Yorkston, K.M.: Speech and pause characteristics following speech rate reduction in hypokinetic dysarthria. J. Commun. Disord. **2**(6), 429–445 (1996)

Harel, B., Cannizzaro, M., Snyder, P.J.: Variability in fundamental frequency during speech in prodromal and incipient Parkinson's disease: a longitudinal case study. Brain Cogn. **56**(1), 24–29 (2004)

Ho, A.K., Iansek, R., Marigliani, C., et al.: Speech impairment in a large sample of patients with Parkinson's disease. Behav. Neurol. **11**(3), 131–137 (1999)

Ho, A.K., Bradshaw, J.L., Iansek, R.: For better or worse: the effect of levodopa on speech in Parkinson's disease. Mov. Disord. **23**(4), 574–580 (2008)

Holmes, R.J., Oates, J.M., Phyland, D.J., Hughes, A.J.: Voice characteristics in the progression of Parkinson's disease. Mov. Disord. **35**(3), 407–418 (2000)

Jiménez-Jiménez, F.J., Gamboa, J., Nieto, A., et al.: Acoustic voice analysis in untreated patients with Parkinson's disease. Park. Relat. Disord. **3**(2), 111–116 (1997)

Karlsson, F., Olofsson, K., Blomstedt, P., et al.: Articulatory closure proficiency in patients with Parkinson's disease following deep brain stimulation of the subthalamic nucleus and caudal zona incerta. J. Speech Lang. Hear. Res. **57**(4), 1178–1190 (2014)

Klawans, H.L.: Individual manifestations of Parkinson's disease after ten or more years of levodopa. Mov. Disord. **1**(3), 187–192 (1986)

Lafargue, G., D'Amico, A., Thobois, S., et al.: The ability to assess muscular force in asymmetrical Parkinson's disease. Cortex **44**(1), 82–89 (2008)

Lang, A.E., Lozano, A.M.: Parkinsons disease: first of two parts. N. Engl. J. Med. **339**(1), 1044–1053 (1998)

Liu, H.M., Tsao, F.M., Kuhl, P.K.: The effect of reduced vowel working space on speech intelligibility in Mandarin-speaking young adults with cerebral palsy. J. Acoust. Soc. Am. **117**(6), 3879–3889 (2005)

Liu, H., Wang, E.Q., Metman, L.V., Larson, C.R.: Vocal responses to perturbations in voice auditory feedback in individuals with Parkinson's disease. PLoS ONE **7**(3), e33629 (2012)

Logemann, J., Fisher, H.: Vocal tract control in Parkinson's disease: phonetic feature analysis of misarticulations. J. Speech Hear. Disord. **46**, 348–352 (1981)

Ludlow, C.L., Connor, N.P., Bassich, C.J.: Speech timing in Parkinson's and Huntington's disease. Brain Lang. **32**(2), 195–214 (1987)

Ma, J.K.-Y.: Lexical tone production by Cantonese speakers with Parkinson's disease. In: Proceedings of INTERSPEECH, pp. 1691–1694, Brighton, UK (2009)

Ma, J.K.-Y., Ciocca, V., Whitehill, T.L.: The perception of intonation questions and statements in Cantonese. J. Acoust. Soc. Am. **129**(2), 1012–1023 (2011)

Ma, J.K.-Y., Hoffmann, R.: Acoustic analysis of intonation in Parkinson's disease. In: Proceedings of INTERSPEECH, pp. 2586–2589, Makuhari, Japan (2010)

Ma, J.K.-Y., Whitehill, T., Cheung, K.S.: Dysprosody and stimulus effects in Cantonese speakers with Parkinson's disease. Int. J. Lang. Commun. Disord. 45(6), 645–655 (2010a)

Ma, J.K.-Y., Whitehill, T.L., So, S.Y.: Intonation contrast in Cantonese speakers with hypokinetic dysarthria associated with Parkinson's disease. J. Speech Lang. Hear. Res. 53(4), 836–849 (2010b)

Mawdsley, C., Gamsu, C.V.: Periodicity of speech in Parkinsonism. Nature 231(5301), 315–316 (1971)

Metter, E.J., Hanson, W.R.: Clinical and acoustical variability in hypokinetic dysarthria. J. Commun. Disord. 19(5), 347–366 (1986)

Miller, N., Andrew, S., Noble, E., Walshe, M.: Changing perceptions of self as a communicator in Parkinson's disease: a longitudinal follow-up study. Disabil. Rehabil. 33(3), 204–210 (2011)

Möbes, J., Joppich, G., Stiebritz, F., et al.: Emotional speech in Parkinson's disease. Mov. Disord. 23(6), 824–829 (2008)

Munson, B.R., Babel, M.: The Phonetics of Sex and Gender (in press)

Nakano, K.K., Zubick, H., Tyler, H.R.: Speech defects of parkinsonian patients: effects of levodopa therapy on speech intelligibility. Neurology 23(8), 865–870 (1973)

Ramig, L.O., Fox, C., Sapir, S.: Speech treatment for Parkinson's disease. Expert Rev. Neurother. 8(2), 299–311 (2008)

Rigaldie, K., Nespoulous, J.L., Vigouroux, N.: Dysprosody in Parkinson's disease: musical scale production and intonation patterns analysis. In: Proceedings of Speech Prosody, Dresden, Germany (2006)

Roy, N., Nissen, S.L., Dromey, C., Sapir, S.: Articulatory changes in muscle tension dysphonia: evidence of vowel space expansion following manual circumlaryngeal therapy. J. Commun. Disord. 42(2), 124–135 (2009)

Rusz, J., Cmejla, R., Ruzickova, H., et al.: Quantitative acoustic measurements for characterization of speech and voice disorders in early untreated Parkinson's disease. J. Acoust. Soc. Am. 129(1), 350–367 (2011)

Sapir, S.: Multiple factors are involved in the dysarthria associated with Parkinson's disease: a review with implications for clinical practice and research. J. Speech Lang. Hear. Res. 57(4), 1330–1343 (2014)

Sapir, S., Ramig, L.O., Spielman, J.L., Fox, C.: Formant centralization ratio (FCR): a proposal for a new acoustic measure of dysarthric speech. J. Speech, Lang. Hear. Res. 53(1), 114–125 (2010)

Skodda, S., Grönheit, W., Schlegel, U.: Intonation and speech rate in Parkinson's disease: general and dynamic aspects and responsiveness to levodopa admission. J. Voice 25(4), 199–205 (2011a)

Skodda, S., Grönheit, W., Schlegel, U.: Impairment of vowel articulation as a possible marker of disease progression in Parkinson's disease. PLoS ONE 7(2), e32132 (2012)

Skodda, S., Rinsche, H., Schlegel, U.: Progression of dysprosody in Parkinson's disease over time: a longitudinal study. Mov. Disord. 24(5), 716–722 (2010a)

Skodda, S., Schlegel, U.: Speech rate and rhythm in Parkinson's disease. Mov. Disord. 23(7), 985–992 (2010)

Skodda, S., Visser, W., Schlegel, U.: Short- and long-term dopaminergic effects on dysarthria in early Parkinson's disease. J. Neural Transm. 117(2), 197–205 (2010b)

Skodda, S., Visser, W., Schlegel, U.: Vowel articulation in Parkinson's disease. J. Voice 25(4), 467–472 (2011b)

Solomon, N.P., Hixon, T.J.: Speech breathing in Parkinson disease. J. Speech Hear. Res. **36**(2), 294–310 (1993)

Solomon, N.P., Robin, D.A.: Perceptions of effort during handgrip and tongue elevation in Parkinson's disease. Park. Relat. Disord. **11**(6), 353–361 (2005)

Turner, G.S., Tjaden, K., Weismer, G.: The influence of speaking rate on vowel space and speech intelligibility for individuals with amyotrophic lateral sclerosis. J. Speech Hear. Res. **38**(5), 1001–1013 (1995)

Viallet, F., Jankowski, L., Purson, A., Teston, B.: L-DOPA effects on speech dysprosody in Parkinson's disease: an acoustic and aerodynamic study. In: Proceedings of International Congress of Parkinson's Disease & Movement Disorders, Chicago, IL, USA (2008)

Weismer, G.: Articulatory characteristics of Parkinsonian dysarthria: segmental and phrase-level timing, spirantization, and glottal-supraglottal coordination. In: McNeil, M., Rosenbeck, J., Aronson, A. (eds.) The Dysarthrias: Physiology, Acoustics, Perception, Management, pp. 101–130. College Hill Press, San Diego (1984)

Weismer, G., Jeng, J.Y., Laures, J.S., et al.: Acoustic and intelligibility characteristics of sentence production in neurogenic speech disorders. Folia Phoniatr. Logop. **53**(1), 1–18 (2001)

Wolfe, V.I., Garvin, J.S., Bacon, M., Waldrop, W.: Speech changes in Parkinson's disease during treatment with L-DOPA. J. Commun. Disord. **8**(3), 271–279 (1975)

Dynamic Acoustic Evidence of Nasalization as a Compensatory Mechanism for Voicing in Spanish Apraxic Speech

Anna K. Marczyk[1,2]([envelope]) [iD], Yohann Meynadier[2], Yulia Gaydina[2], and Maria-Josep Solé[3] [iD]

[1] Brain and Language Research Institute, 5 av. Pasteur,
13100 Aix-en-Provence, France
anna.marczyk@lpl-aix.fr
[2] Aix-Marseille Université, CNRS, LPL, Aix-en-Provence, France
[3] Universitat Autònoma de Barcelona, Barcelona, Spain

Abstract. This paper is concerned with the phonetic realization of the voicing contrast by two Spanish speakers with surgery-related apraxia of speech and two matched control speakers. Specifically, it examines whether speakers with AOS, widely reported to have a deficit in laryngeal control, use nasal leak as a compensatory mechanism aimed at facilitating the initiation of voicing in word-initial stops. The results show that the two apraxic speakers produced prevoicing in /b d g/ in only one third of the cases (correctly identified as 'voiced'). In these cases, however, they exhibited significantly longer prevoicing than control subjects, and this longer voiced portion was closely related to a longer nasal murmur. These results shed light on the compensation strategies used by apraxic subjects to achieve voicing. Differences in the intensity patterns of nasal and voiced stops indicate that apraxic speakers control the timing of velopharyngeal gesture, suggesting that apraxia is a selective impairment.

Keywords: Nasality · Voicing · Apraxia of speech · Acoustics

1 Introduction

Apraxia of speech (AOS) is a motor speech disorder of neurological origin that selectively affects phonetic encoding processes [1–4] and results in distortions of the sound shape of words. Phonetic investigations of apraxic speech have yielded evidence for impaired laryngeal control, timing and coordination with supralaryngeal articulators [5]. This deficit results in frequent devoicing errors, a hallmark of AOS across languages, especially in phrase- or word-initial consonants. The initiation of voicing, however, is difficult not only for speech-impaired speakers. Indeed, aerodynamic conditions for phrase-initial stops may require additional motor adjustments to favor

© Springer Nature Switzerland AG 2018
Q. Fang et al. (Eds.): ISSP 2017, LNAI 10733, pp. 225–236, 2018.
https://doi.org/10.1007/978-3-030-00126-1_20

voicing. In 'true voicing'[1] languages, such as Spanish, nasal leakage has been reported to be a common facilitatory mechanism associated with voicing initiation and maintenance [6]. Nasal leak helps to evacuate the air behind an oral closure and maintain the transglottal pressure differential necessary to initiate glottal pulsing. In this paper, we seek to determine whether two Spanish speakers with AOS use this strategy to enhance voicing perception and analyze how their phonetic implementation differs from that of healthy speakers.

Articulatorily, nasality involves synchronizing the velic movement with the oral release. The difference between nasals and prenasalized voiced stops lies in the fact that the velic closure occurs almost simultaneously to the oral release in nasals and before the oral release in prenasalized stops. These differences are captured by acoustic metrics of duration and amplitude [7, 8]. Thus, in this paper we will combine analyses in the temporal and intensity domains.

The synchronization of articulatory events may be problematic for speakers with apraxia. Yet, while difficulties in the coordination of the laryngeal and oral gestures have been widely reported, the phonetic realization of nasal consonants is relatively unimpaired [9]. We hypothesize that apraxic subjects may employ nasal leak to promote voicing and we predict that, when an initial voiced stop is successfully produced, analyses will reveal acoustic traces of compensatory activity (e.g., a longer nasal murmur during the consonant closure).

While the mechanism described above may prove useful to initiate voicing in stop consonants, apraxic speakers must finely control the timing of velopharyngeal closure and thus avoid productions that fall into the 'nasal' category. Our analyses aim at comparing nasals and voiced stops to determine whether there are differences in their phonetic realization between control and apraxic speakers.

Our ultimate goal is to gain a better understanding of how the timing deficit in apraxia may affect the phonetic realization of voiced stops involving prenasalization.

2   Methodology

2.1   Participants

Two Spanish female speakers aged 34 and 37, both right-handed, diagnosed with apraxia of speech related to high-grade glioma resection, and two matched controls took part in this study. The tumors were revealed by epileptic seizures and transient expressive aphasia. A pre-surgical fMRI assessment showed scarce bilateral activation during language tasks with a strong left lateralization of the expressive language

---

[1] It is well known that Spanish /b d g/ are typically produced with voicing lead ('pre-voicing') such that the onset of vocal fold vibration precedes the release of the stop, resulting in negative VOT values, while /p t k/ are produced with a near-simultaneous release and onset of laryngeal vibration, resulting in VOT values that are approximately zero [10, 11]. Voiceless stops are phonetically realized as stops, i.e., with a complete oral closure, in all contexts. By contrast, voiced /b d g/ are realized as stops utterance-initially, after a nasal, or after [l] in the case of /d/, and systematically realized as approximants in all other contexts, for example, between vowels or after a continuant.

functions, primarily involving the pars opercularis area of the inferior frontal gyrus (BA44) and the premotor cortex, for both patients. For speaker 1 the growing tumor mass induced a displacement and deformation of the anterior segment of the left arcuate fasciculus. Both patients underwent an awake craniotomy guided by direct brain-mapping. The lesions were well-circumscribed and restricted to the cortical area with little infiltration. Initially mute, both patients underwent intensive speech therapy and evolved from an acute stage towards Broca's-type aphasia and, finally, apraxia of speech. Clinical assessment 12 months post-surgery revealed no deficit in naming, auditory comprehension, reading, writing or repetition abilities, apraxia of speech being the only postoperative symptom at the time of testing. Their speech was characteristically slow and slurred, with syllable-by-syllable articulation, frequent phonetic distortions and errors surfacing as phoneme substitutions, especially devoicing of word-initial stops. Overall these symptoms were compatible with those observed in stroke-related AOS.

2.2 Stimuli

Acoustic data were obtained for isolated words elicited in word reading and repetition tasks. We used material compiled for a larger study on the phonetic realization of Spanish consonants [12], from which we selected bi- and tri-syllabic words with word-initial nasals /m n/, voiceless /p t k/ and voiced stops /b d g/. The number of tokens analyzed comprised 22 nasals, 296 voiceless stops and 251 voiced stops for each group of speakers. The consonants of interest were always followed by a non-high vowel /e a o/.

2.3 Analysis Procedure

Recordings of the elicited consonant productions were independently transcribed by two phoneticians and classified as on-target nasals, voiceless stops, voiced stops, or errors (i.e., 'voiced stops' identified as 'voiceless'). No instances of voiced stops heard (i.e., transcribed) as nasals were found. Voiced stops were also categorized as 'pre-nasalized' if they exhibited nasal murmur, a weak formant structure and increased amplitude of voicing on the acoustic records.

Because our data were limited to acoustic output only, it was not straightforward to infer whether nasal leak had occurred and, if so, whether or not it was used in combination with other adjustments aimed at maintaining a low oral pressure for voicing such as cavity enlargement. Nonetheless, the acoustic records obtained showed that prevoiced /b d g / in Spanish could present (i) a strong low frequency resonance at about 200–250 Hz which is the main resonance of the nasal cavity, and (ii) a weak formant structure that resembled vowel formants, with large antiresonances caused by the interaction between oral and nasal cavity resonances. These realizations were considered to involve nasal leak and are illustrated in Fig. 1 for control and apraxic speakers.

Furthermore, we obtained aerodynamic data for the same control and apraxic subjects for a larger corpus with the EVA2 data acquisition system [13]. The aerodynamic data revealed that nasal leak exhibited the acoustic characteristics described

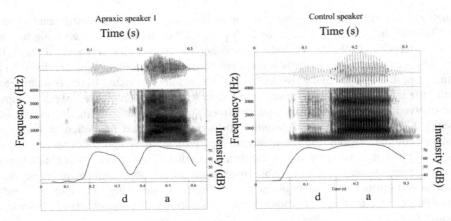

Fig. 1. Waveform, spectrogram (0–4 kHz) and amplitude (dB) of the syllable /'da/ in the word *dato* ('information') produced by an apraxic (left) and a control speaker (right), showing voicing and prenasalization.

above. This is illustrated in Fig. 2, which shows nasal leak throughout the stop closure for utterance-initial /b/ and the acoustic result. Thus we feel confident that nasal leak may be inferred from the acoustic signal.

Analyses in the Temporal Domain. Three acoustic parameters were measured: voicing lead, voicing lag and nasal murmur duration during stop closure. For VOT analysis, in cases where periodicity was uninterrupted from the onset of glottal pulsing to the noise burst generated at the constriction (whether due to passive tissue expansion only or its co-occurrence with nasal leak), it was measured as negative VOT (Fig. 1, right). If voicing was initiated but ceased after a few tens of ms (passive devoicing), presumably due to failure to maintain nasal leak, two measurements were made. First, as per the revisited definition of VOT [14], such cases were regarded as negative VOT. However, given that the duration of devoicing can be informative of underlying pathological processes, the second set of measurements considered each element separately: voiced portion with nasal murmur, devoiced portion and positive VOT (Fig. 1, left). Analyses were carried out by means of linear regression intended to detect differing patterns in the distribution of these parameters between control and apraxic subjects.

Analyses in the Intensity Domain. Visual inspection of the intensity contours showed that data points constituting the intensity trajectories tended to display a curvilinear shape (Fig. 1). Using Praat [15], we extracted the intensity listing at every 0.01s for every nasal, voiced and voiceless stop. In order to adjust for variation in segment duration, temporal normalization was applied throughout the process (i.e., the time coordinate was always between 0 and 1).

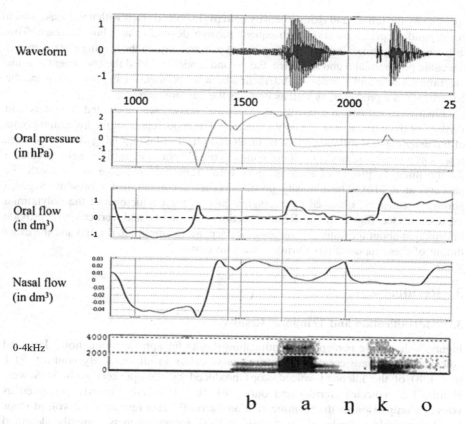

Fig. 2. Waveform, intraoral pressure, oral flow, nasal flow and spectrogram (0–4 kHz) for Spanish *banco* ('bank') by an apraxic speaker. In the oral flow channel, the level of no airflow (0 dm³) is indicated by a dashed line. Long vertical lines mark onset of voicing and stop release.

The curve fitting procedure was then applied using the cubic polynomial equation

$$y(x) = ax3 + bx2 + cx + d \qquad (1)$$

where y corresponds to the intensity value (in dB) and x to a point in time, in order to transform the data into a set of four coefficients, each of them carrying information relative to different aspects of the curve. The estimated coefficients were saved as variables for statistical analyses.

Our first hypothesis predicted differences in intensity values (y) between nasals and voiced stops across groups. Since the coefficient d corresponds to the intensity value at consonant onset (if $x = 0$, all the other coefficients are set to 0 and y equals d), it was entered as the dependent variable in a linear regression mixed model with phonemic category (nasals vs. voiced stops), group (apraxic vs. control) and their interaction as predictors, and items as a random factor.

The second linear regression mixed model was aimed at testing the hypothesis relative to differences in intensity dynamics across phonemic categories and groups, specifically the falling and rising intensity patterns in voiced stops with nasal leak, which correspond to prenasalization, subsequent passive devoicing and burst release. This difference is expressed through the covariance pattern between the leading coefficients of the cubic polynomial a and b. Hence, the second model included the coefficient a as the outcome variable and a three-way interaction between phonemic category, group and the covariate b as a predictor, as well as items as the random intercept.

Finally, to support the results of the model above, we calculated the roots and vertices of the derivative of the cubic function. The roots represent the inflection points of intensity curve oscillations, that is, onset and offset of falling/rising slopes. Vertices represent intensity values (y) at these points in time. Roots and vertices were calculated for the intensity profile of each category using functions available in the rootSolve package in R [16]. The intensity profile denotes a contour that represents a given phonemic category, obtained by averaging each of the coefficients of the polynomial expression (see Fig. 6 in the Results section). The roots and vertices will provide information about the values of increase or decrease in amplitude (in dB) and the exact timing of these modulations during consonant closure.

3   Results

3.1   Identification and Temporal Results

Figure 3 shows the perceptual identification results for apraxic productions. Intended voiceless stops were correctly identified as voiceless in all cases. By contrast, 71% (n = 178) of the intended voiced stops produced by the speakers with AOS were identified as voiceless (error) and only 29% (n = 73) were correctly perceived as voiced. Comparison of the identification and acoustic data revealed that voiced stops heard as voiceless displayed short positive VOT values whereas correctly identified voiced stops showed prevoicing. There was no significant difference in VOT values between on-target voiceless stops and voiced stops identified as voiceless (errors) [$F_{(1)}$ = .7, p = .400; M = 27 ms (SE = 1.44) for voiceless vs. M = 29 ms (SE = 1.89) for voiced stops]. In the control group, voiceless stops were always correctly identified and only six word-initial /b d g / (2.4%) were identified as devoiced (error).

Turning to temporal parameters, the first set of analyses was aimed at comparing VOT values for apraxic and control speakers in correctly perceived productions to identify potential compensatory mechanisms in the former. For correctly identified voiced and voiceless stops, the analyses revealed a significant main effect of phonemic category (voiced vs. voiceless) [$F_{(1)}$ = 2038, p = .000], no significant effect of group, and a significant interaction between phonemic category and group [$F_{(1)}$ = 325.34, p = .000], indicating differences in the phonetic implementation of voicing across populations. As shown in Fig. 4, this difference was due to voiced stops, which showed significantly longer voicing lead in apraxic speech as compared to the speech of healthy controls [M = −177 ms (SE = 4.75) vs. M = −67 ms (SE = 5.11)].

Further analyses, limited to correctly identified voiced stops in the apraxic group (n = 73), showed that tokens classified as prenasalized (48 cases out of 73) exhibited overall 28 ms longer voicing lead than voiced stops without nasal leak [M = −179 ms

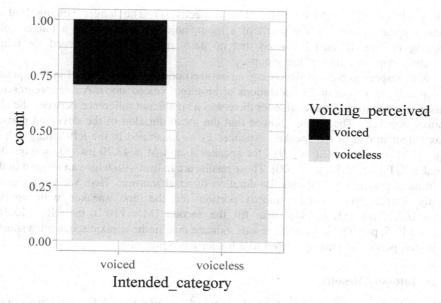

Fig. 3. Relative frequencies of intended word-initial voiced and voiceless stops produced by apraxic speakers, as perceived by listeners. Voiceless stops, n = 296; voiced stops, n = 251.

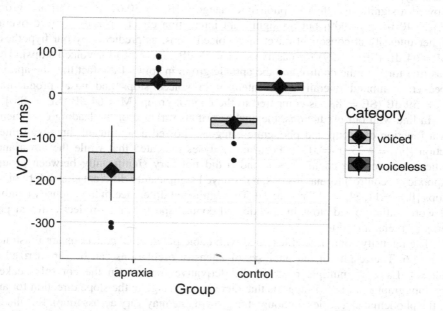

Fig. 4. Distribution of VOT values for correctly identified voiced and voiceless stops in AOS and control groups.

(SE = 10) vs. M = −151 ms (SE = 16), respectively]. The longer voicing lead in apraxic speakers may be the result of a motor adjustment involving a longer velic opening gesture. It must be noted that no nasal murmur was observed for initial voiceless stops in either of the groups.

With respect to passive devoicing, a devoiced portion was observed in the apraxic group only, in 19 out of 73 realizations of on-target voiced stops. A linear regression was performed to determine whether there was a significant difference between the two apraxic speakers. The results showed that the mean duration of the devoiced portion was longer in one of the speakers (speaker 2), as illustrated in the left-hand graph in Fig. 5 [M = 7.43 ms (SE = 3.91) for speaker 1 and M = 43.70 ms (SE = 6.66) for speaker 2, $F_{(1)}$ = 22.54, p = .000]. These results are in line with what can be seen in the right-hand graph, which shows the duration of nasal murmur. Here we see a significantly longer prenasalized (voiced) portion for the first speaker with apraxia [M = 167.17 ms (SE = 8.43)] than for the second [M = 110.10 ms (SE = 15.36), $F_{(1)}$ = 11.75, p = .001]. Thus the results indicate that, in the apraxic speakers, a shorter devoiced portion is closely related to a longer nasal murmur.

3.2 Intensity Results

The results for intensity were subjected to analysis in order to look for differences in the intensity envelope across groups and phonemic categories in accordance with our initial hypotheses. A linear regression analysis of intensity values at consonant onset showed a significant effect of phonemic category [$F_{(1)}$ = 30.62, p = .000] and group [$F_{(1)}$ = 40.14, p = .000] but no significant interaction effects. Nasals showed overall higher intensity at consonant onset than voiced stops, as predicted by our hypothesis [M = 64 dB (SE = 1.09) for nasals vs. M = 58 dB (SE = .56) for voiced stops]. This was true for both the control and the apraxic group in spite of the fact that the apraxic speakers exhibited overall lower intensity in voiced stops and nasal productions [M = 59 dB (SE = .88) as compared to the control group (M = 64 dB (SE = .85)].

In line with our predictions, the analyses of co-variance of the leading coefficients as a function of group and phonemic category showed a significant three-way interaction [$F_{(1)}$ = 4.35, p = .037]. Post-hoc analyses indicated that while the covariance pattern between the coefficients a and b did not vary significantly between groups (apraxic vs. control) for nasal stops, it displayed significantly different slopes for voiced stops [β = −.11, SE = .02, p = .000]. The significant difference in the intensity contour of word-initial voiced stops in apraxic and control speakers may reflect different patterns of prenasalization.

The intensity contour profiles fitted with cubic polynomial equations are illustrated in Fig. 6. The results of our analyses of intensity oscillations can be summarized as follows. Lack of multiple roots of the derivative function in the control speakers (bottom graphs in Fig. 6) suggests that there is no change in the slope direction for any of the phonemic categories (although the growth rate may vary across time). Results for the apraxic group (top graphs) show that nasal stops exhibit a similar pattern across speakers with a slight drop in intensity (1 dB) at half the closure. For voiced stops, this decline is greater (2–4 dB) and occurs earlier. These results are provided in Table 1 and illustrated in Fig. 6.

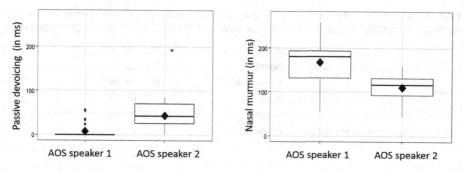

Fig. 5. Distribution of passive devoicing (left) and nasal murmur (right) for the two apraxic speakers.

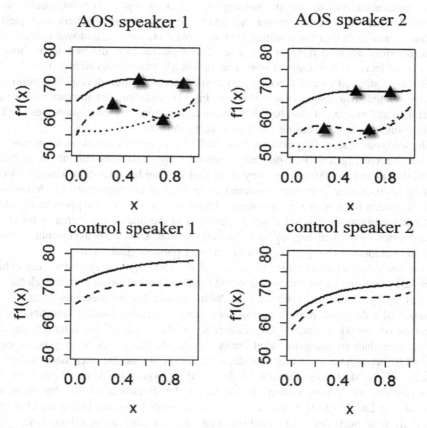

Fig. 6. Intensity profiles fitted with a three-term polynomial equation for two apraxic (top) and two control speakers (bottom). Solid lines: nasals; dashed lines: voiced stops; dotted lines (for apraxic subjects only): voiced stops identified as voiceless (errors). Triangles represent vertices/roots.

Table 1. Summary of the results for intensity oscillation analyses. x^1 and x^2 represent roots of the function, diff(dB) represents the intensity values at x^1 and x^2 (i.e., vertices of the function).

	Phonemic category	x^1.	x^2	diff (dB)
AOS speaker 1	Nasal	0.5	0.9	72–71
	Voiced stop	0.3	0.7	64–60
AOS speaker 2	Nasal	0.5	0.8	69–68
	Voiced stop	0.2	0.6	58–56
Control speakers	No multiple roots			

4   Discussion and Conclusion

This study provides preliminary evidence for the use of motor adjustments—in this case prenasalization—to initiate voicing by speakers with a neurologically-based speech disorder. Our results reveal that when the two apraxic speakers attempted to produce Spanish /b d g/ the resulting output showed prevoicing, and was perceived as 'voiced', only about a third of the time. These productions differed from those in control subjects in that they exhibited significantly longer prevoicing (Fig. 4). The results also showed that for both apraxic speakers a shorter devoiced portion was closely related to a longer nasal murmur (Fig. 5) revealing the interaction between nasal leak and voicing. Taken together, these data give us an idea of strategies used by apraxic subjects to compensate for their voicing deficits.

In sum, our results suggest that nasal leak allows apraxic speakers to initiate and sustain the voicing lead in word-initial voiced stops, and thus enhance a listener's perception of voicing. Moreover, they show that voicing lead duration (negative VOT) may be an important parameter to detect these kind of compensatory mechanisms in apraxic speech in 'true voice' languages. These results also lend support to the claim that apraxic impairment is selective. Differences in the intensity patterns of nasal and voiced stops (shown in Fig. 6) seem to indicate that apraxic speakers control rather well the velopharyngeal gesture as a function of phonological contrast.

On the other hand, our results suggest that, while apraxic speakers may exhibit prenasalization, they also exhibit disturbed temporal coordination between glottal and supraglottal gestures, as evidenced by both temporal measurements—that is, the presence of a devoiced portion during the consonant closure (widely described in the literature on apraxic speech)—and differences in the slopes of the intensity contour. Still, rather than focusing on what apraxic speakers cannot do, it may be of equal interest to investigate what they *can* do, as well as examine the interplay between the deficit and the strategies that can be deployed to bypass it. Along these lines, our understanding of 'compensation' is similar to the definition offered by Simmons-Mackie and Damico [17] as any adaptive behavior that has the following characteristics: (i) it is purposeful and goal-oriented, (ii) it occurs as novel or functionally expanded behavior, (iii) it is context-sensitive and (iv) it is specific to the speaker.

Finally, the observations reported here add new evidence to the existing body of descriptions of symptoms associated with apraxia of speech. While devoicing errors in AOS have been frequently studied, especially in relation to the origin of these errors along the stages of phonetic and phonemic encoding processes, to our best knowledge this is the first study that addresses the interaction between nasal and laryngeal gestures in this disorder. Moreover, studies on surgery-related speech disorders are scarce. Severe cognitive, aphasic or apraxic impairments in such etiologies (as opposed to stroke) are relatively rare, due to the fact that a slow-growing tumor allows the cognitive function to reorganize [18–22]. In the case of the patients studied in this paper, moderate impairment persisted a year after surgery.

Given its limitations, this work should be regarded as a preliminary exploration. Future analyses need to include a larger number of speakers, for example, or incorporate not just acoustic but also aerodynamic data. It would also be of considerable interest to make comparisons across languages that use different ranges of VOT. Such research would help us to build a truly complete understanding of the production patterns of voiced stops and compensatory gestures in apraxia of speech.

Acknowledgements. We are grateful to Tim Mahrt for his help with R scripting, Thierry Legou for his assistance with polynomial equations and María Machuca for her help with data transcription. This research has been supported by grants ANR-11-LABX-0036 (BLRI), ANR-11-IDEX-0001- (A*MIDEX), FFI2013-46354-P and FFI2017-84479-P, Ministry of Science and Innovation, Spain.

References

1. Ballard, K.J., Granier, J.P., Robin, D.A.: Understanding the nature of apraxia of speech: theory, analysis, and treatment. Aphasiology **14**(10), 969–995 (2000)
2. Code, C.: Models, theories and heuristics in apraxia of speech. Clin. Linguist. Phon. **12**(1), 47–65 (1998)
3. Maas, E., Gutiérrez, K., Ballard, K.J.: Phonological encoding in apraxia of speech and aphasia. Aphasiology **28**(1), 25–48 (2014)
4. Ziegler, W.: Psycholinguistic and motor theories of apraxia of speech. Semin. Speech Lang. **23**, 231–243 (2002)
5. Blumstein, S.E., Cooper, W.E., Goodglass, H., Statlender, S., Gottlieb, J.: Production deficits in aphasia: a voice-onset time analysis. Brain Lang. **9**(2), 153–170 (1980)
6. Solé, M.-J.: Articulatory adjustments in initial voiced stops in Spanish, French and English. J. Phon. **66**, 217–241 (2018)
7. Burton, M., Blumstein, S.E., Stevens, K.N.: A phonetic analysis of prenasalized stops in Moru. J. Phon. **20**, 127–142 (1992)
8. Kong, E.J., Syrika, A., Edwards, J.R.: Voiced stop prenasalization in two dialects of Greek. J. Acoust. Soc. Am. **132**(5), 3439–3452 (2012)
9. Kurowski, K., Blumstein, S.E., Palumbo, C.L., Waldstein, R., Burton, M.: Nasal consonant production in Broca's and Wernicke's aphasics: speech deficits and neuroanatomical correlates. Brain Lang. **100**(3), 262–275 (2008)
10. Borzone de Manrique, A., Gurlekian, J.: Rasgos acústicos de las conso-nantes oclusivas españolas. Rev. Fonoaudiológica **26**, 326–330 (1980)
11. Quilis, A.: Fonética acústica de la lengua española. Gredos, Madrid (1981)
12. Baqué, L., Estrada, M., Nespoulous, J.-L., Le Besnerais, M., Rosas, A., Marczyk, A.: Corpus léxico del proyecto COGNIFON. Barcelona: Unpublished internal document (2008)

13. Ghio, A., Teston, B.: Evaluation of the acoustic and aerodynamic constraints of a pneumotachograph for speech and voice studies. In: International Conference on Voice Physiology and Biomechanics, Marseille, pp. 55–58 (2004)
14. Abramson, A.S., Whalen, D.H.: Voice Onset Time (VOT) at 50: theoretical and practical issues in measuring voicing distinctions. J. Phon. **63**, 75–86 (2017)
15. Boersma, P., Weenink, D.: Praat: doing phonetics by computer (2017). http://www.praat.org/
16. Soetaert, K., Herman, P.M.J.: A Practical Guide to Ecological Modelling. Using R as a Simulation Platform. Springer, New York (2009). doi:https://doi.org/10.1007/978-1-4020-8624-3
17. Simmons-Mackie, N., Damico, J.S.: Reformulating the definition of compensatory strategies in aphasia. Aphasiology **11**(8), 761–781 (1997)
18. Davie, G.L., Hutcheson, K.A., Barringer, D.A., Weinberg, J.S., Lewin, J.S.: Aphasia in patients after brain tumour resection. Aphasiology **23**(9), 1196–1206 (2009)
19. Duffau, H., et al.: Usefulness of intraoperative electrical subcortical mapping during surgery for low-grade gliomas located within eloquent brain regions: functional results in a consecutive series of 103 patients. J. Neurosurg. **98**, 764–778 (2003)
20. Ojemann, J., Miller, J., Silbergeld, D.: Preserved function in brain invaded by tumor. Neurosurgery **39**, 253–259 (1996)
21. Plaza, M., Gatignol, P., Leroy, M., Duffau, H.: Speaking without Broca's area after tumor resection. Neurocase **15**(4), 294–310 (2009)
22. Satoer, D., Visch-Brink, E., Dirven, C., Vincent, A.: Glioma surgery in eloquent areas: can we preserve cognition? Acta Neurochir. **158**, 35–50 (2016)

Author Index

Printed in the United States
by Bookmasters

Printed in the United States
By Bookmasters